An Introduction to Atomic Absorption Spectroscopy

An Introduction to Atomic Absorption Spectroscopy

A Self-teaching Approach

L. Ebdon
Reader in Analytical Chemistry
Plymouth Polytechnic, Devon, UK

LONDON • PHILADELPHIA • RHEINE

Heyden & Son Ltd, Spectrum House, Hillview Gardens, London NW4 2JQ, UK
Heyden & Son Inc., 247 South 41st Street, Philadelphia, PA 19104, USA
Heyden & Son GmbH, Devesburgstrasse 6, 4440 Rheine, West Germany

---

British Library Cataloguing in Publication Data

Ebdon, L.
   An introduction to atomic absorption spectroscopy.
   1. Atomic absorption spectroscopy
   I. Title
   535.8'4   QC454.A8

ISBN 0-85501-714-7

© Heyden & Son Ltd, 1982

All Rights Reserved. No part of this publication may be reproduced, stored in a retrieval system, or transmitted, in any form or by any means, electronic, mechanical, photocopying, recording or otherwise, without the prior permission of the copyright holder.

Photoset by Paston Press, Norwich
Printed in Great Britain by Mackays of Chatham Ltd

CONTENTS

PREFACE ix
ACKNOWLEDGEMENTS xi
Aims xiii
Prior Knowledge xiii
Introduction and Study Time-table xiii

1 **INTRODUCTION** 1
  1.1 Historical 1
  1.2 Basic Instrumentation 3
  1.3 Optics for Spectrometers 4
  1.4 Detectors 8

2 **FLAME SPECTROSCOPY** 11
  2.1 Flame Structure 11
  2.2 Flame Temperatures 12
  2.3 Flame Gas Mixtures 14
  2.4 Sample Introduction and Sample Atomization . . . 16
  2.5 Burner Design 20
  2.6 Flame Spectra 22
  2.7 Broadening 24

3 **FLAME EMISSION SPECTROSCOPY** 27
  3.1 Theory 27
  3.2 Instrumentation 29
  3.3 Other Sources for Atomic Emission Spectroscopy . 32
    3.3.1 Introduction 32
    3.3.2 Solid sampling emission sources 34
    3.3.3 Plasma sources 36

4 **FLAME ATOMIC ABSORPTION SPECTROSCOPY** . . 42
  4.1 Theory 42
  4.2 Instrumentation 45
    4.2.1 Sources 45

	4.2.2 The most popular atom cells: flames	48
	4.2.3 Instrument design	49
	4.2.4 Read-out systems	55
	4.2.5 Sensitivity and limit of detection	56

5 FLAME ATOMIC FLUORESCENCE SPECTROSCOPY ... 58
5.1 Theory 58
5.2 Instrumentation 59
    5.2.1 Sources 59
    5.2.2 Flames 61
    5.2.3 Instrument design 62

6 INTERFERENCES AND ERRORS IN FLAME SPECTROSCOPY 63
6.1 Sample Pre-treatment Errors 63
6.2 Operator Errors 64
6.3 Instrument Errors 65
6.4 Interferences 66
    6.4.1 Spectral interferences 66
    6.4.2 Ionization interferences 67
    6.4.3 Chemical interferences 67

7 THE PRACTICE OF FLAME SPECTROSCOPY 71
7.1 Applications 73
    7.1.1 Clinical, food and organic samples 73
    7.1.2 Agricultural samples 73
    7.1.3 Water and effluents 75
    7.1.4 Geochemical and mineralogical samples . . . 75
    7.1.5 Metals 76
    7.1.6 Solvent extraction of trace metals 76
7.2 Comparison of the Analytical Utility of Atomic Emission Spectroscopy, Atomic Absorption Spectroscopy and Atomic Fluorescence Spectroscopy 77

8 MODIFICATIONS TO FLAME SPECTROSCOPY 80
8.1 The Limitations of Conventional Flame Cells 80
8.2 Modified Flame Cells 81
    8.2.1 Pulse nebulization 81
    8.2.2 Branched uptake capillaries 81
    8.2.3 Kahn sampling boat 82
    8.2.4 Delves sampling cup 82
8.3 Further Arguments about Flame Atom Cells 83

CONTENTS

9 MERCURY DETERMINATION 86
   9.1 The Reduction–Aeration Method 86
   9.2 Cold-vapour Atomic Absorption Spectroscopy 87
   9.3 Cold-vapour Atomic Fluorescence Spectroscopy 88

10 HYDRIDE GENERATION 90
   10.1 Chemistry 90
   10.2 Instrumentation 91

11 ELECTROTHERMAL ATOMIZATION 92
   11.1 Historical Development 92
   11.2 Heated Graphite Furnace Atomizers 95
   11.3 Filament and Mini-Furnace Graphite Atomizers 98
   11.4 Recent Developments in Electrothermal
       Atomizer Designs 99
       11.4.1 Control of furnace temperature 99
       11.4.2 Tube dimensions 100
       11.4.3 Isothermal operation 100
   11.5 Atomization Mechanisms 101
       11.5.1 Thermodynamic considerations 101
       11.5.2 Kinetic considerations 102
   11.6 Interferences 103
       11.6.1 Physical interferences 103
       11.6.2 Background absorption 104
       11.6.3 Memory effects 104
       11.6.4 Chemical interferences 104
   11.7 Applications 106
   11.8 The Relative Merits of Electrothermal
       Atomization 107
       11.8.1 Advantages of electrothermal atomization . . . 107
       11.8.2 Disadvantages of electrothermal atomization . . 109

APPENDICES 110
  Appendix A: Revision Questions 110
  Appendix B: Practical Exercises 114
    Calculations 114
    Safety procedures 117
    Experiments 119
  Appendix C: References and Bibliography 130

INDEX 134

PREFACE

Atomic absorption spectroscopy offers an excellent example of the need for continuing study and training. No other instrumental analytical technique has ever shown such rapid growth in applications, yet the theory and methodology of atomic absorption are but poorly covered in many basic chemistry courses. It is hard to escape the conclusion that this sad state of affairs arises not only from rapid growth, but also because of a certain unprofessional aloofness on the part of many teachers, possibly affronted at the wide industrial popularity of a technique so elegantly simple and undemanding of spectroscopic theory, and a certain jealously by advocates of superseded analytical techniques.

Rapid technological development and the increasingly volatile nature of employment in the scientific world ensure that few, if any, can expect to leave school or college equipped with a range of skills and learning which will serve all our working life. Indeed, part of the current debate about full-time education is the extent to which vocational studies should appear on the curriculum. We can certainly agree that an aim of such education should be the equipping of students with the abilities necessary for self-teaching after graduation.

Traditionally, those wishing to acquire new knowledge in mid-career or at the start of a new career have turned either to evening classes or correspondence courses. In advanced vocational subjects, the evening class has developed, via day release, into part-time day and evening courses, typified in the UK by several M.Sc. courses in Instrumental Chemical Analysis and many post-HC courses in Advanced Analytical Chemistry (the so-called LRSC courses). The less-generous economic climate now prevailing has placed considerable difficulties in the way of those seeking day release to go on such courses, or attempting to attend more intensive short courses on single techniques. For several years, I was involved in the teaching of atomic absorption spectroscopy on the relatively-successful part-time M.Sc. course in Instrumental Chemical Analysis at Sheffield City Polytechnic. When we came to review the course prior to the periodic resubmission to the Council for National Academic Awards, we considered this dilemma. There was clearly a growing need for the course, yet potential students were finding it increasingly difficult to obtain the necessary half-day release. We were aware that one of the most spectacular educational successes of the past decade has been the Open University. The major features of their

PREFACE

method of study are not the much-publicized televised components, but the books of study material. If this approach could be successful at the undergraduate level, it seemed likely that it should be applicable at the postgraduate level. The decision was taken to teach half the traditional lecture programme by individualized learning methods. This book sprang from that decision. The material which I wrote to teach the part of the syllabus covering atomic absorption, emission and fluorescence spectroscopy in flame and electrothermal atom cells was so well received by the students, that I was encouraged to publish the course for a wider public.

This book, while it is based upon that original self-learning material, has been expanded and amended in several ways. Obviously, I have tried to ensure that the contents have been brought completely up to date, even in those areas where developments are being made rapidly. The book now assumes no advanced prior knowledge and is as self-contained a course as is possible without dilution of the basic aim to teach atomic absorption spectroscopy. An attempt has been made to retain a succinct style of writing with keywords highlighted to aid revision. At the suggestion of the students who found them helpful, I have included more questions and I hope this will aid readers who have to study this book independently.

The contents follow the structure that students have told me is most useful. After a brief historical introduction, common elements of instrumentation are considered, followed by a detailed discussion of the flame as an atom cell. The following chapters consider, in turn, flame emission, flame atomic absorption and flame atomic fluorescence spectroscopy. In view of the very wide interest in it as the technique of choice for trace metal analysis, atomic absorption is dealt with in most detail. In each case, the exposition develops from basic theory, through practical instrumentation to real application. Succeeding chapters deal in depth with practical aspects such as interferences and applications, before attention is directed to the recent, significant growth in the use of atom cells other than flames. The final chapter looks at electrothermal atomization in considerable detail, concluding with a survey of the relative merits of this approach.

I hope that the reader will find this book a real contribution to the literature of analytical chemistry. Several eminent workers in the field have commented to me on the lack of an introductory book suitable for self-teaching the newcomer to atomic absorption spectroscopy. While I should be proud to think that the old 'AA hand' can find something new and of value here, I shall derive most satisfaction from helping the student or analyst who is meeting an atomic absorption spectrometer for the first time. I hope that, like me, they will find this useful technique elegant in its simplicity and versatile in its capability, and come to find the practice of analytical atomic spectroscopy profoundly satisfying.

Les Ebdon
Plymouth
July 1981

ACKNOWLEDGEMENTS

I acknowledge with thanks the following for permission to reproduce copyright material: W. J. Price for Fig. 1.7, taken from *Spectrochemical Analysis by Atomic Absorption*, Heyden & Son, London (1979); G. F. Kirkbright, A. Semb, T. S. West and Maxwell Scientific Inc., New York, for Figs 2.9 and 2.10, taken from *Talanta* **14**, 1014 (1967); W. B. Barnett, J. W. Vollmer and S. M. DeNuzzo and the Perkin–Elmer Corporation, Norwalk, Connecticut for Fig. 4.3, taken from *Atomic Absorption Newsletter*, **15**, 33 (1976); H. T. Delves and the Royal Society of Chemistry, London, for Fig. 8.1, taken from the *Analyst* **95**, 431 (1970); B. V. L'vov and Pergamon Press, Oxford, for Fig. 11.1, taken from *Spectrochimica Acta* **24B**, 55 (1969); H. Massman and Pergamon Press, Oxford, for Fig. 11.2, taken from *Spectrochimica Acta* **23B**, 217 (1968); P. J. Whiteside and Pye-Unicam Ltd., for Fig. 11.4.1; C. W. Fuller and the Royal Society of Chemistry, London, for Figs 11.4.2 and 11.4.3, taken from *Electrothermal Atomization for Atomic Absorption Spectrometry*, Chemical Society, London (1977); and K. C. Thompson, D. R. Thomerson and the Royal Society of Chemistry, London, for Table 10.1, which originally appeared in the *Analyst* **99**, 595 (1974).

Figures 2.2 and 2.3 are adapted from Lewis and Van Elbe, *Combustion, Flames and Explosions of Gases*, Academic Press, New York (1961). Thanks are also due to Pye–Unicam Ltd., Instrumentation Laboratory, and Varian Techtron, from whose sales literature Figs 4.1, 4.8 and Table 7.1, respectively, are taken or adapted.

Thanks are also due to all those who have encouraged me in this undertaking, particulary those students who first recommended publication and those, including my research students, who made helpful comments on presentation. I am grateful for the support of former colleagues at Sheffield City Polytechnic, especially Dr. M. Goldstein, Head of Chemistry, for first suggesting that my lecture material might be presented in this way. No author should omit mention of the inspiration he has received from his own teachers. In my case, mention must be made of Professor G. F. Kirkbright and Professor T. S. West.

Finally, my largest debt is to my family for their forbearance during the preparation of this book, most notably my wife Judith who typed the original manuscript and its many drafts.

AIMS

To introduce and develop a knowledge of the theory, instrumentation and practice of atomic absorption, emission and fluorescence spectroscopy in flame and electrothermal atom cells.

To discuss the applications of these techniques to a range of typical problems in analytical chemistry.

To compare the relative merits of the above techniques in analytical situations.

To introduce an understanding of currently developing themes in the field and enable the student to follow current advances through the means of the scientific literature.

To enable the student to develop and use precise and accurate analytical methods for a range of determinations using the above techniques.

PRIOR KNOWLEDGE

An attempt is made in this book to approach analytical atomic spectroscopy without requiring extensive prior knowledge. An understanding of the structure of the atom and basic spectroscopy is required, but this is not assumed to be at graduate level; for most of the book, an understanding at the level of the Bohr theory will suffice. It is assumed that some basic instrumentation is known, and no detailed discussion of optical or standard electrical components is given.

INTRODUCTION AND SUGGESTED STUDY TIME-TABLE

The learning material is designed to replace traditional learning systems, such as lectures, as adequately as possible. While therefore it covers a definable 'course' it is, like lectures, designed to be read in conjunction with supplementary texts. Background reading in the original and current literature is very valuable.

By setting aside an evening a week for study, the reader may complete this book in five weeks. A suggested study time-table is given below.

Week 1	Chapters 1 and 2
Week 2	Chapters 3, 4 and 5
Week 3	Chapters 6 and 7
Week 4	Chapters 8, 9 and 10
Week 5	Chapter 11

Questions are interspersed with the material. You should satisfy yourself that you can answer a question from the material above it before proceeding.

1 INTRODUCTION

1.1 HISTORICAL

Spectroscopy is generally considered to have started in 1666, with **Newton's** discovery of the solar spectrum. **Wollaston** repeated Newton's experiment and in 1802 reported that the sun's spectrum was intersected by a number of dark lines. **Fraunhofer** investigated these lines—Fraunhofer lines—further, and in 1823 was able to measure their wavelengths.

Early workers had noted the colours imparted to diffusion flames of alcohol by metallic salts, but detailed study of these colours awaited the development of the premixed air–coal gas flame by **Bunsen**. In 1859, **Kirchhoff** showed that these colours arose from line spectra due to elements and not compounds. He also showed that their wavelengths corresponded to those of the Fraunhofer lines. Kirchhoff and Fraunhofer had been observing atomic emission and atomic absorption, respectively.

Atomic absorption spectroscopy (AAS), atomic emission spectroscopy (AES) and later atomic fluorescence spectroscopy (AFS) then became more associated with an exciting period in astronomy and fundamental atomic physics. Atomic emission spectroscopy (AES) was the first to re-enter the field of analytical chemistry, initially in arc and spark spectrography and then through the work of **Lunegardh**, who in 1928 demonstrated AES in an air–acetylene flame using a pneumatic nebulizer. He applied this system to agricultural analysis.

The term **flame emission spectroscopy** is applied to the measurements of light emitted from flames by chemical species after the absorption of energy as heat or as chemical energy (i.e. chemiluminescence). An older term is **flame photometry**. If only the emission from atoms is observed, the term **atomic emission spectroscopy** is preferred.

Atomic absorption spectroscopy is the term used when the radiation absorbed by atoms is measured. The application of AAS to analytical problems was considerably delayed because of the apparent need for very high resolution to make quantitative measurements. In 1953, **Walsh** brilliantly overcame this obstacle by use of a line source, an idea pursued independently by **Alkemade**, his work being published in 1955.

The re-emission of radiation from atoms which have absorbed light is

AN INTRODUCTION TO ATOMIC ABSORPTION SPECTROSCOPY

termed **atomic fluorescence**. In 1962, **Alkemade** was the first to suggest that AFS had analytical potential, which was demonstrated in 1964 by **Winefordner**.

These three types of spectroscopy are summarized diagrammatically in Fig. 1.1.

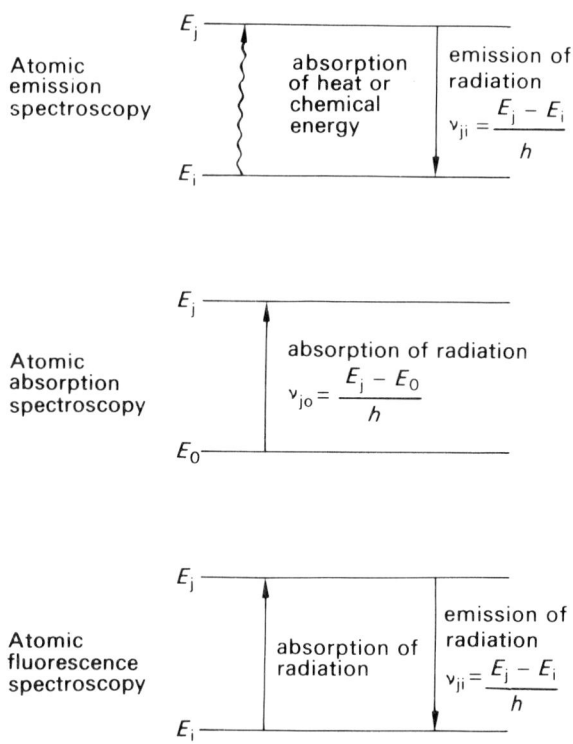

Fig. 1.1   Summary of AES, AAS and AFS.

The horizontal lines represent different energy levels in an atom. E_0 is the term used for the lowest energy level, which is referred to as the **ground state**. In flames, atoms are most commonly encountered in the ground state and therefore all **practical absorption** measurements originate from atoms in the ground state, as do virtually all practical fluorescence measurements. E_i and E_j refer to other energy levels, E_j being higher (greater energy) than E_i. A solid vertical line refers to a **transition** involving the absorption or emission of energy as radiation. The wavy line refers to a **non-radiative transition**. The energy of the radiation absorbed or emitted is quantized according to **Planck's equation**, $E = h\nu$, where h is Planck's constant, ν is the frequency of the radiation and E is the energy difference between the two energy levels in the atom. The frequency is related to **wavelength** by the formula λ (wavelength) $= c$ (speed of light)$/\nu$.

INTRODUCTION

Q. What is the fundamental difference between AES and AFS?

1.2 BASIC INSTRUMENTATION

Figure 1.2 shows the basic instrumentation necessary for each technique. At this stage, we will define the component where the atoms are produced and viewed as the 'atom cell'. Much of what follows will explain what we mean by this term. In atomic emission spectroscopy, the atoms are excited in the atom cell also, but for atomic absorption spectroscopy and atomic fluorescence spectroscopy, an external light source is used to excite the ground state atoms. In atomic absorption spectroscopy, the source is viewed directly and the attenuation of radiation measured. In atomic fluorescence spectroscopy, the source is not viewed directly, but the re-emitted radiation is measured.

Current instrumentation usually uses a **diffraction grating** as the dispersive element and a **photomultiplier** as the detector. Rapid-scanning photoelectric detectors, such as Vidicon tubes, have been used in some research instruments.

A variety of **read-out systems** are in use. These include analogue meters, often with the capability of reading linearly in absorbance, chart recorders, paper tape print-out, digital display and several microprocessor-controlled

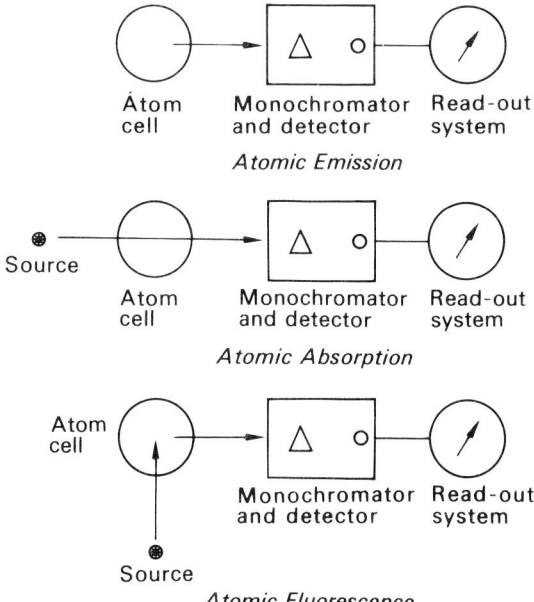

Fig. 1.2 Basic instrumental systems used in analytical atomic spectroscopy.

displays. With the latter, calibrations may be performed automatically, and data on concentration, standard deviation etc. presented on a visual display unit.

As may be seen, most of these components are common to a wide variety of instrumentation, and in the next two sections some typical arrangements will be outlined.

Q. Explain the different geometries used for AAS and AFS.

1.3 OPTICS FOR SPECTROMETERS

To exploit the full potential of atomic spectroscopy, it is usually necessary to isolate a **narrow band of wavelengths**. As will be seen, this is particularly important for atomic absorption spectrometry. This isolation may be achieved by the use of **filters** or by **geometrical dispersion**. Dispersion is generally preferred, although filters may be used in some simple emission instruments.

Coloured glass filters which absorb most radiation, but allow some wavelengths to be transmitted, are widely available. Unfortunately, their **transmission windows** are quite large (e.g. 40 nm), which severely limits their use. Filters which function on the interference principle, i.e. **interference filters**, transmit much smaller bands of wavelength (e.g. 10 nm) and are more useful, but expensive. They can be made by depositing a very thin, **semi-transparent** film of silver on a glass or quartz plate, covering this film with a very thin layer of a **transparent material** such as magnesium fluoride and then with another **semi-transparent layer** of silver. As each silver film reflects about half of the radiation that strikes it, some of the light will be **repeatedly reflected** before it is transmitted. The emergent rays will reinforce each other only for radiation with a wavelength which is an **exact multiple** of twice the distance between the two silver films. For all other radiation, the distance travelled by the light rays in the filter is not an exact multiple of their wavelength, and so the beams interfere destructively. The transmitted radiation is **more intense** than that from a coloured glass filter, but consists of multiple orders. The unwanted orders can be removed by making the base plate, or a cover plate, absorptive in the correct region.

A **monochromator** is an instrument that can isolate a narrow range of wavelengths (e.g. 1–0.01 nm) anywhere in a comparatively **wide spectral range** (for atomic absorption spectroscopy, typically 190–900 nm). The better resolution and the ability to select any desired wavelength make monochromators the preferred means of wavelength isolation for atomic absorption spectrometry.

INTRODUCTION

Although most modern instruments use **replica diffraction gratings** to achieve dispersion, many early instruments used **prism monochromators**. As many of the most analytically useful atomic absorption lines occur in the ultraviolet, prisms made of silica (e.g. fused quartz) were used. The **60° prism** traditionally used in spectrographic instruments can be replaced in a more **compact** and **economical** way by a **30° prism** with its back surface silvered. This so-called **Littrow design** is shown in Fig. 1.3. The passage of radiation through the quartz in both directions will also correct for any **birefringence**. The use of a **concave mirror** rather than a lens to focus the light on the prism and on the **exit slit** enables the coverage of a greater **optical range** and **eliminates chromatic aberration**, a significant problem with lenses.

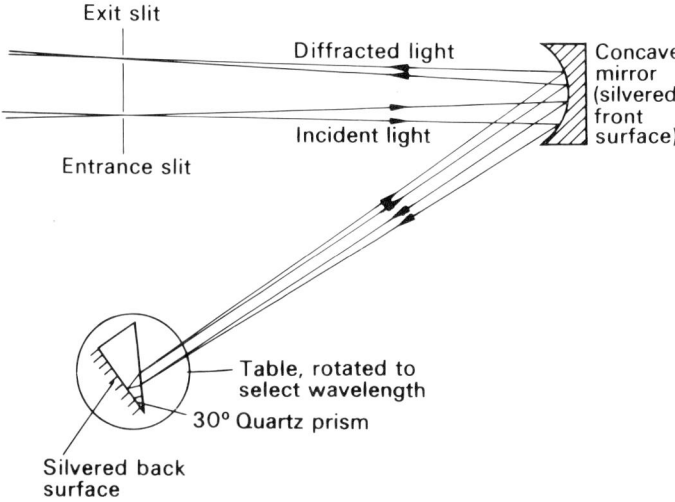

Fig. 1.3 Littrow-mounted prism monochromator.

The **dispersion** given by a prism is high in the ultraviolet, but **decreases** rapidly **as wavelength increases**. Gratings give a dispersion which is essentially **constant** throughout the spectrum, depending instead on the number of **rulings** per unit width and the **focal length** of the monochromator. The **energy** transmitted by a prism may be greater, as it is not split into different orders and there are usually few **stray light** or **ghost** problems (faint false lines on each side of the true spectral line). As **gratings** have improved, stray light and ghost problems have been diminished, and **blazing** the grating enables most of the energy to be thrown into the first order. The **cost** of replica gratings now compares favourably with that of prisms. **Holographic gratings**, which can be cut into a photosensitive resin using the interference fringes of two coherent lasers, offer considerable improvements in performance, and with modern technology they can also be blazed economically. Hence, the use of gratings is becoming increasingly popular.

The theory of a diffraction grating involves **Huygens' principle** of secondary wavelets and is really beyond our scope. A simplified discussion of arguments for a transmission grating may be helpful; this can easily also be applied to a reflection grating. A **transmission grating** consists of a transparent plate with a great many lines ruled upon it (e.g. 20 000 lines cm^{-1}). These act like regularly spaced narrow slits. From these, the light will emerge as a series of intersecting wavelets like a series of **semi-circles** emanating from **each slit**, i.e. as if each slit were itself a source of radiation. In a short time, the wave systems will recombine to form the original wavefront, the so-called **zero-order**. More interestingly, at a **series of angles**, θ, to the grating, there will be **constructive interference** for light of a given wavelength. The position of these diffraction wavefronts is given by the **grating equation**:

$$n\lambda = d \sin \theta$$

where λ is the wavelength, d is the space between the rulings on the grating, and n is an integer (0, 1, 2, 3 ...) called the **order**. Thus, a beam of polychromatic radiation is diffracted into a series of spectra symmetrically located on either side of the normal to the grating. It should be noted that for any given θ there will be several different wavelengths, albeit in different orders. This **overlapping** problem is reduced by **blazing** the grating to a level which generally gives rise to no problems, and may be eliminated by the use of filters or a preliminary **order sorting** prism.

Figure 1.4 shows schematically a **blazed reflection grating**. A ray incident at an angle α will be reflected from the groove face at an angle β, such that $\alpha + \phi = \beta - \phi$, where ϕ is the **angle of blaze**. Such a grating, also known as an echelette grating, is highly efficient in the diffraction of wavelengths close to those for which specular reflection occurs. The wavelength at which specular reflection and first-order diffraction coincide is called the blaze wavelength, λ_B. Energy is concentrated in the first order at λ_B, at half λ_B in the second order, and so on.

There are other types of special gratings. An **echelle grating** is ruled with step-shaped rulings a few hundred times wider than the average wavelength to be studied. This is in contrast to the general arrangement where the line spacing is similar to the average wavelength. The echelle is used, however, at orders, n, of 100 or more and is thus capable of **considerable dispersion**. The manufacture of precision gratings is very demanding and **replica** gratings are **cast** in plastic from an original.

A grating may be mounted in a monochromator in several ways. One method uses the **Littrow** mounting, shown in Fig. 1.3. Another arrangement, the **Ebert** mounting (Fig. 1.5), uses a large spherical mirror to collimate and focus the beam. **Czerny and Turner** suggested replacing the large,

INTRODUCTION

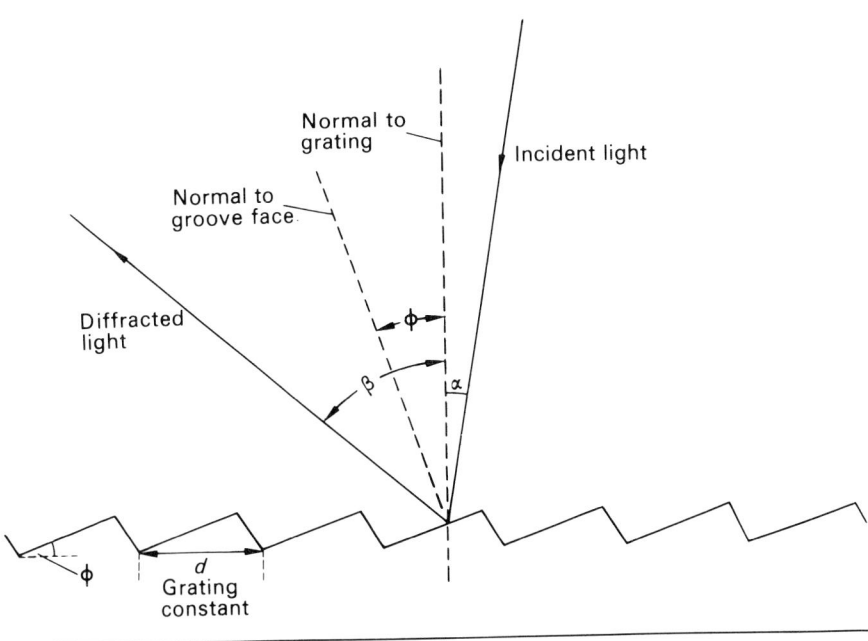

Fig. 1.4 Schematic diagram of a blazed reflection grating.

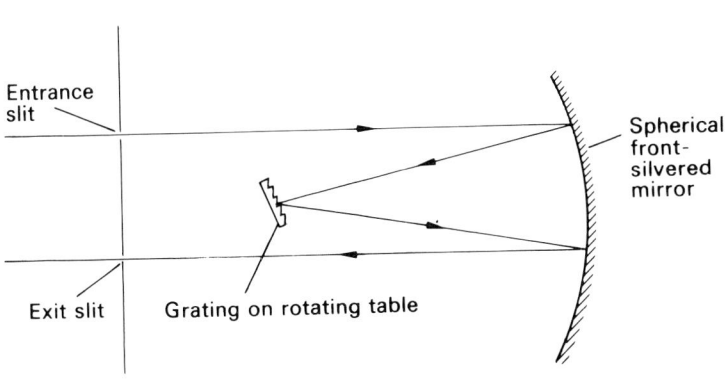

Fig. 1.5 Ebert mounting for a reflection grating.

expensive Ebert mirror with two small, spherical mirrors mounted symmetrically, as shown in Fig. 1.6. The Czerny–Turner mounting is very popular for atomic absorption spectrometers as it combines economy with relative freedom from aberrations.

AN INTRODUCTION TO ATOMIC ABSORPTION SPECTROSCOPY

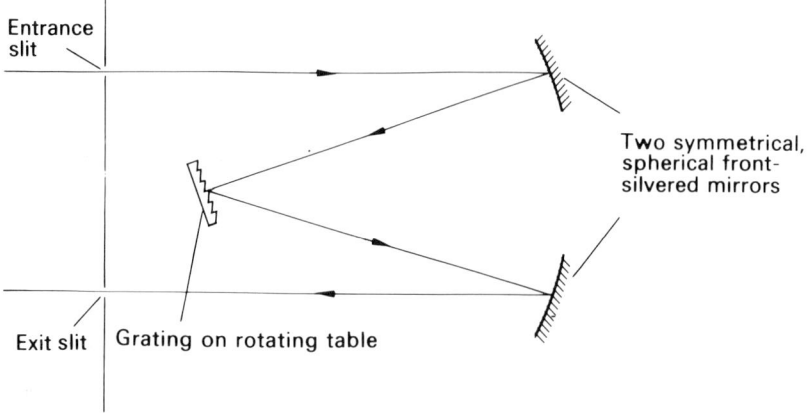

Fig. 1.6   Czerny–Turner monochromator.

Q. Which is the preferred method of wavelength selection for AAS?

Q. What limits the use of coloured glass filters?

Q. How does an interference filter work?

Q. What are the relative advantages and disadvantages of prisms versus gratings?

Q. What is the advantage of blazing a grating?

Q. What is the difference between a Czerny–Turner and an Ebert mounting?

1.4   DETECTORS

Photomultiplier tubes are now in almost universal use as detectors of spectral radiation, although photographic plates and photocells were used in the past. A photomultiplier tube consists of a **photosensitive cathode** and a series of **dynodes** at successively more **positive potentials**, culminating in an anode. The whole device, enclosed in a **vacuum**, has a **window** to allow the radiation into the tube. When photons strike the cathode, **electrons** are ejected which are then **accelerated** down the dynode chain. Each electron impacting upon the first dynode causes a number of **secondary electrons** to be ejected which are focused on to the next dynode, and the process repeats itself. In this way, four secondary electrons generated from the first dynode

INTRODUCTION

in a chain of 12 dynodes would produce 17 million electrons per photon. For a typical duration of 5 ns, this would produce an anode current of the order of 1 mA.

The correct choice of the photosensitive material used to coat the cathode is important. Usually it is a semi-conductive material containing an **alkali** metal. Different cathode materials have different **response curves**, as may be seen from Fig. 1.7. It can be seen that caesium–antimony cathodes operate well up to about 550 nm, but their response at the red end (e.g. for the potassium or rubidium lines at 766 and 780 nm, respectively) is poor. For these longer wavelengths, a trialkali cathode, sodium–potassium–antimony with a trace of caesium, may be preferred. The recently introduced gallium arsenide cathodes offer a reasonably uniform response over a wide range (even up to the caesium line at 852 nm). Response at the low wavelength end (e.g. the arsenic line at 193.7 nm) is often dependent upon

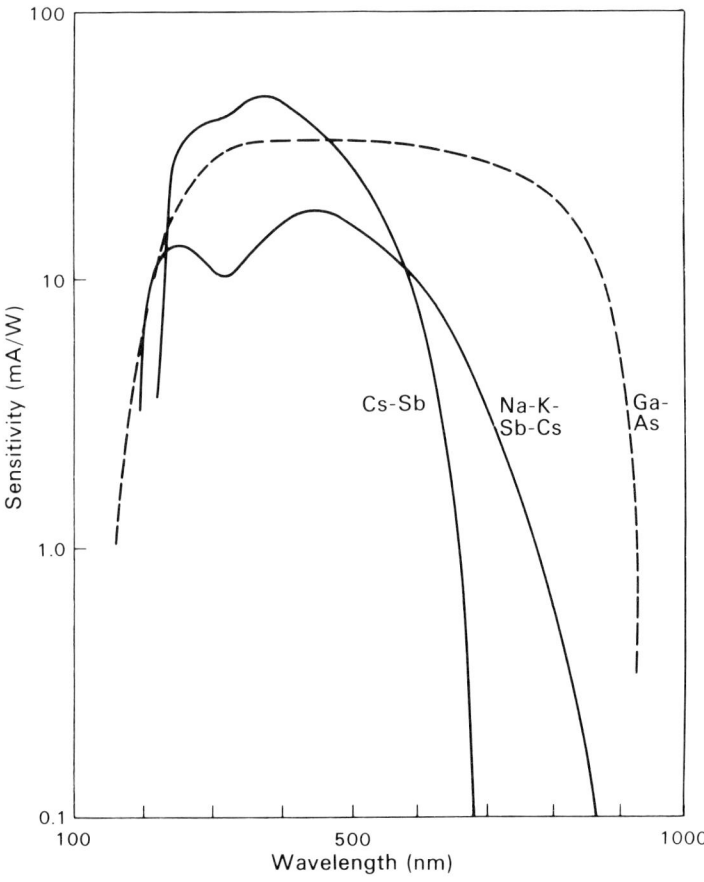

Fig. 1.7 Photomultiplier sensitivity curves.

AN INTRODUCTION TO ATOMIC ABSORPTION SPECTROSCOPY

the material used for the window (either placed **'end-on'**, which is easier to make, or **'side-on'**) and its transmission at low wavelengths.

For certain applications, a photomultiplier which only operates in the ultraviolet region may be preferred. Such a photomultiplier does not respond to daylight and is known as **solar blind**. Ordinary photomultiplier tubes can be damaged if operated without protection from daylight.

It is important that a photomultiplier gives low noise and a low **dark current**, i.e. a low background signal in the absence of photons. Obviously, increasing the dynode voltage, and hence the amplification, tends to increase these unwanted signals. Thus, the high voltage supply (commonly up to 1000 V) must be highly stabilized and is only increased as necessary. This is usually performed using the knob marked 'gain', which is connected to a potentiometer. In modern instruments, automatic gain control may be provided to do this.

A discussion of typical **read-out** systems for atomic absorption spectrometers is given later (section 4.2.4). One feature common to many spectroscopic instruments is **modulation**. If the analytical signal is **coded** by modulation (e.g. by applying an AC supply to the spectral source for atomic absorption or atomic fluorescence or by interrupting the beam by a synchronized rotating sector), and the amplifier is **tuned** to the modulation frequency, the analytical signal can be distinguished from spurious, noise signals. Thus, an **AC amplifier** can **distinguish** between the signal of interest and other signals. The most convenient modulation **frequency** to use is the normal mains frequency, 50–60 Hz. Below this frequency, much noise, for example **flame flicker**, will break through as it has a low-frequency AC component. Higher frequencies, e.g. 285 Hz, which also has the advantage of not being a mains harmonic, are more difficult to achieve, but offer better discrimination against **noise**. Fast transient signals may also be followed with less distortion at higher frequencies.

Q. How is amplification achieved in a photomultiplier tube?

Q. What cathode material would you prefer to use to measure lithium atomic emission at the 671 nm line?

Q. What is dark current?

Q. What is modulation and why is it so widely used?

2 FLAME SPECTROSCOPY

Although the popularity of electrothermal atomization (ETA) has increased, flames are still used for over 90% of all analytical applications. As the **most useful atom cell**, they are worthy of detailed study. Flames are extensively discussed by Kirkbright and Sargent (see Appendix C for full details on all references).

2.1 FLAME STRUCTURE

Figure 2.1 shows the structure of a typical **pre-mixed flame**. (In the past, non-premixed or turbulent flames were also used in direct-injection burners, but they are now recognized as less suitable.) Pre-mixed gases are heated in the **pre-heating zone**, where their temperature is raised exponentially until it reaches the **ignition temperature**. Surrounding the pre-heating zone is the **primary reaction zone**, where the most energetic reactions take place.

The primary reaction zone is a hollow cone-like zone, only 10^{-5}–10^{-4} m thick. The actual **shape of the cone** is determined largely by the velocity distribution of the gas mixture leaving the burner. While the velocity of the gases at the burner walls is virtually zero, it reaches a maximum in the centre. The rounding at the top is caused, in part, by thermal expansion of the gases, which also produces a back pressure which distorts the base of the cone, causing some overhang of the burner. The cone elongates as the gas flow is increased. If this flow is increased so much that the gas velocity exceeds the burning velocity, the flame will **'lift-off'**. If the flow is decreased so that the reverse occurs, the flame may **'strike-back'** with possibly explosive effect.

The primary reaction zone is so thin that thermodynamic equilibrium cannot possibly be established in it and the partially combusted gases and the **flame radicals** (e.g. $\cdot OH$, $\cdot H$, $\cdot C_2$, $\cdot CH$ and $\cdot CN$), which propagate the flame, pass into the **inter-conal zone**. Equilibrium is quickly established here as radicals combine. It is usually regarded as the hottest part of the flame and the most favoured for analytical spectrometry.

The hot, partly combusted gases then come into contact with oxygen from the air and the final flame products are formed. This occurs in what is known as the **secondary reaction zone** or diffusion zone.

AN INTRODUCTION TO ATOMIC ABSORPTION SPECTROSCOPY

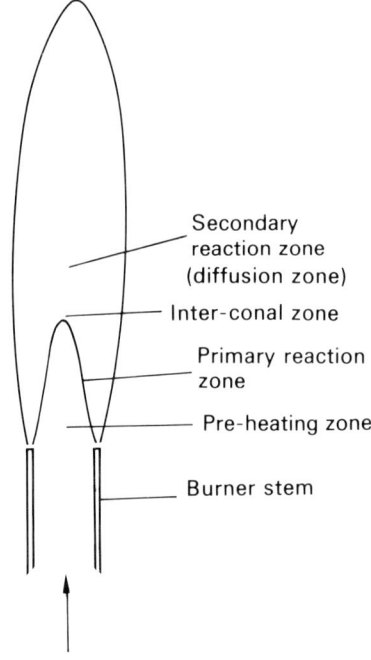

Fig. 2.1 A premixed (or laminar) flame.

The description above refers to pre-mixed gas flames with **laminar** (i.e. non-turbulent) flow of the gas mixture to the flame.

Q. Describe and explain the shape of the primary reaction zone.

Q. How can flames be prevented from striking-back?

Q. Why does a kettle boil fastest when the tips of the blue cones of the Bunsen flame are immediately below its base?

2.2 FLAME TEMPERATURES

Various approaches to measuring flame temperature are well described in Gaydon's book on Flames (see Appendix C). The best methods are spectroscopic rather than those which use thermocouples. The sodium **line reversal method** is perhaps the easiest. Sodium is added to the flame and the sodium D lines viewed against a bright continuum source (e.g. a hot carbon tube). When the flame is cooler than the source, the lines appear dark because of absorption. When the flame is hotter than the tube, the bright lines stand out in emission. The current to the tube, which will have been pre-calibrated for temperature readings by viewing the tube with an optical pyrometer, is adjusted until the lines cannot be seen. At this reversal point, the flame and tube temperature should be equal.

Other methods, based upon **two lines**, may be used. Two-line methods may be used in absorption, emission or fluorescence. The signal is measured at the lines obtained when metal atoms are sprayed into the flame. Provided there is no self-absorption and the transition probabilities of the lines are known accurately, the flame temperature can be calculated from the ratio of line intensities using the **Maxwell–Boltzmann distribution** (see section 3.1).

Flame temperature varies from one part of the flame to another, as indicated in section 2.1. Figures 2.2 and 2.3 show this effect for a stoichiometric and a fuel-rich flame.

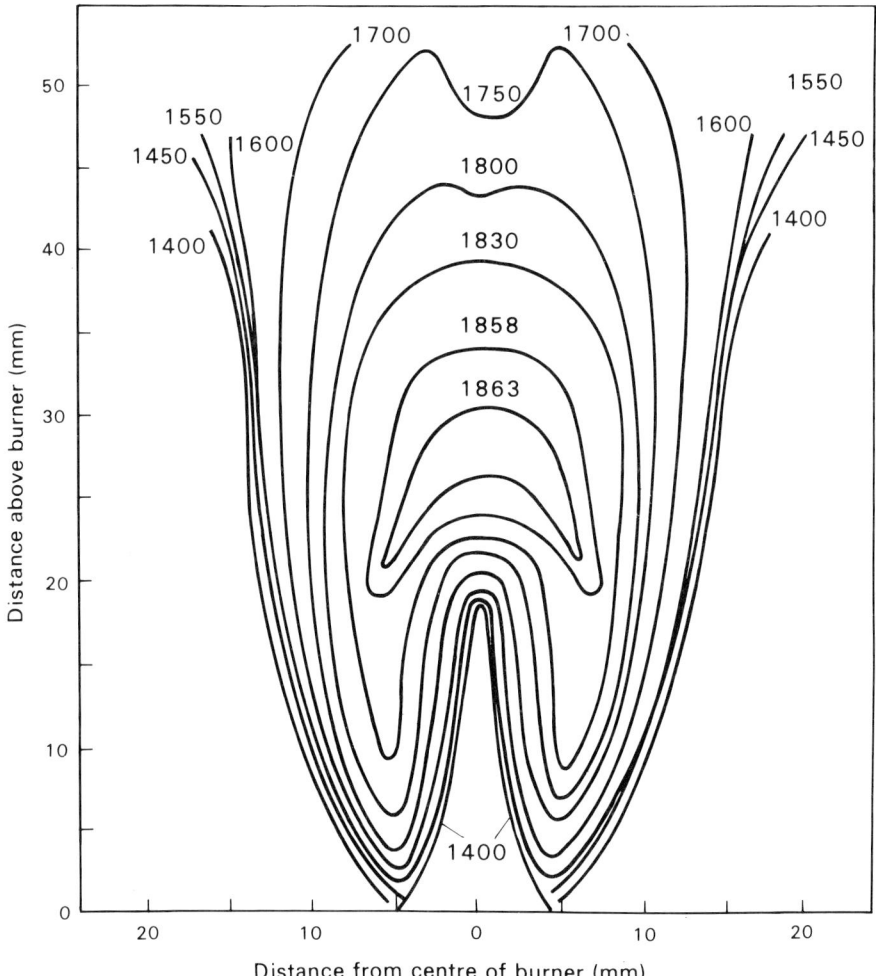

Fig. 2.2 Temperature distribution in a near stoichiometric premixed air–natural gas flame. All temperatures in °C.

AN INTRODUCTION TO ATOMIC ABSORPTION SPECTROSCOPY

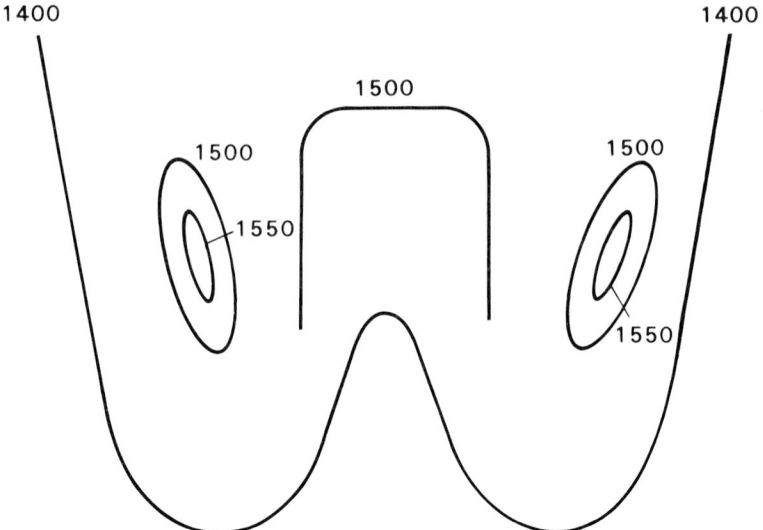

Fig. 2.3 Comparative figure (with Fig. 2.2) showing a fuel-rich flame temperature distribution.

Q. List some ways in which flame temperature may be measured.

Q. Why does line reversal occur?

2.3 FLAME GAS MIXTURES

Table 2.1 lists some characteristics of the most popular pre-mixed flames used for analytical spectrometry.

Table 2.1

Oxidant	Fuel	Flow rates for stoichiometric flame (l min^{-1})		Maximum burning velocity (cm s^{-1})	Approximate experimental temperature (K)
		Oxidant	Fuel		
Air	Propane	8	0.4	45	2200[a,b]
Air	Hydrogen	8	6	320	2300[a,b]
Air	Acetylene	8	1.4	160	2500[a,b,c]
Nitrous oxide	Acetylene	10	4	285	3150[a,c]

[a] A. G. Gaydon and H. G. Wolfhard, *Flames: Their Structure, Radiation and Temperature*, Chapman & Hall, London (1960), p. 304.
[b] J. B. Willis, *Appl. Opt.* **7**, 1295 (1968).
[c] K. M. Aldous, B. W. Bailey and J. M. Rankin, *Anal. Chem.* **44**, 191 (1972).

These values should only be taken as indicative. The figures for the burning velocity and temperature show that different burners need to be used with different flames.

The **air–town-gas** and **air–propane** flames are rarely used nowadays, as they are cool and offer insufficient atomization efficiency. They are, however, easy to handle. The **air–hydrogen** flame finds special use in atomic fluorescence because of the low fluorescence-quenching cross-section of hydrogen, often further improved by diluting (and cooling) the flame with argon.

The **air–acetylene** flame is the most widely used flame. It is stable, simple to operate and produces sufficient atomization to enable good sensitivity and freedom from inter-element interferences for many elements. It is not only necessary for the flame to atomize the analyte, but also to break down any refractory compounds which might react with or physically entrap the analyte. Atomization, as we shall see, occurs both because of the high enthalpy and **temperature** of the flame, and through **chemical** effects. Thus, increasing the oxygen content of the flame above the approximately 20% normally present in air, while raising the flame temperature, does not necessarily enhance atomization, because more refractory oxides may be produced. Making the flame more fuel-rich lowers the temperature, but, by making the flame more reducing, increases the atomization of elements such as molybdenum and aluminium.

The **nitrous oxide–acetylene** flame is both **hot and reducing**. A characteristic red, inter-conal zone is obtained under slightly fuel-rich conditions. This **'red feather'** is due to emission by the **cyanogen radical**. This radical is a very efficient scavenger for oxygen, thus pulling equilibria such as

$$TiO \rightleftharpoons Ti + \tfrac{1}{2}O_2$$

over to the right. This appears to be a vital addition to the high temperature which also promotes dissociation. Amongst those elements which are best determined in nitrous oxide–acetylene are Al, B, Ba, Be, Mo, Nb, Re, Sc, Si, Ta, Ti, V, W, Zr, the lanthanoids and actinoids. The nitrous oxide–acetylene flame must be operated more carefully than the air–acetylene flame. For safety reasons, an air–acetylene flame is lit first, made very fuel-rich and then the air switched to nitrous oxide by a two-way valve. The flame is shut down by the reverse procedure. The nitrous oxide–acetylene flame can usually be run without problems, provided that it is never run fuel-lean and carbon deposits not allowed to build up. Any deposits should be cleaned away when the flame has been extinguished.

Q. Why are different burners needed for different flames?

Q. What are the advantages and disadvantages of the air–propane flame?

AN INTRODUCTION TO ATOMIC ABSORPTION SPECTROSCOPY

Q. Why is the air–acetylene flame so popular for AAS?

Q. What causes the red feather observed in the nitrous oxide–acetylene flame?

Q. What advantages does the nitrous oxide–acetylene flame offer in AAS?

2.4 SAMPLE INTRODUCTION AND SAMPLE ATOMIZATION

So far, we have no analyte atoms in the atom cell! This is usually achieved in the following manner, although some alternative ways are considered in Chapter 8. A more detailed description is given by Kirkbright and Sargent (see Appendix C).

Figure 2.4 shows a typical **pneumatic nebulization** system for a premixed flame. The sample is sucked up a plastic **capillary tube**. In the type of **concentric nebulizer** illustrated here, the sample liquid is surrounded by the oxidant gas as it emerges from the capillary. The high velocity of this gas, as it issues from the tiny annular orifice, creates a pressure drop which sucks up, draws out and 'shatters' the liquid into very tiny **droplets**. This

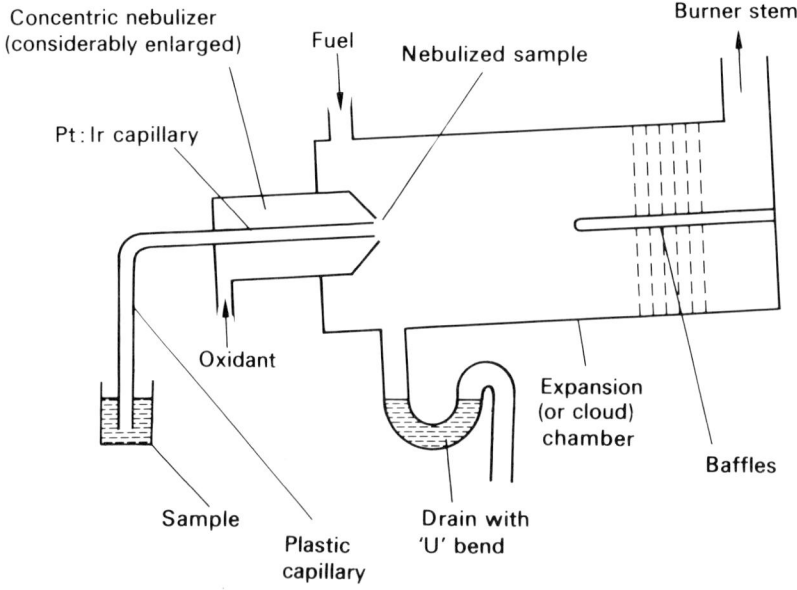

Fig. 2.4 Schematic diagram of a concentric pneumatic nebulizer system for a premixed burner.

FLAME SPECTROSCOPY

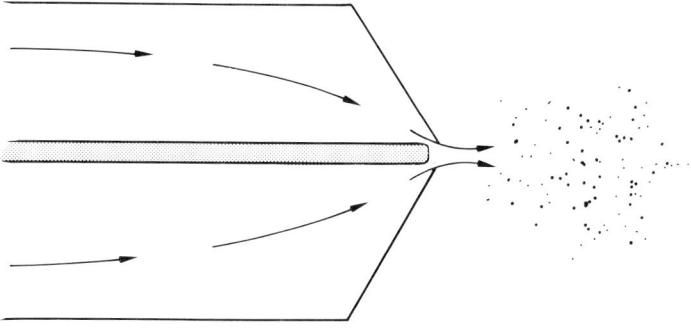

Fig. 2.5 The venturi effect.

phenomenon is known as the **venturi effect** and is illustrated in Fig. 2.5.

The nebulizer capillary position may be **adjustable** on a screw thread to enable optimization of sample uptake and drop size. Alternatively or additionally, an **impact bead** may be placed in the path of the initial aerosol to provide a secondary fragmentation and so improve the efficiency of nebulization. Such a device is illustrated in Fig. 2.6.

The material of the nebulizer must be highly **corrosion resistant**. Commonly, the plastic capillary is fixed to a platinum–iridium alloy (90:10) capillary, mounted in stainless steel gas supply inlets. The impact bead is sometimes made of a similar alloy or smooth borosilicate glass.

The aerosol then passes along the plastic **expansion chamber**. Large droplets collect on the walls of the chamber and, to ensure that only the smallest particles reach the flame, **spoilers** or **baffles** may be placed in the path of the

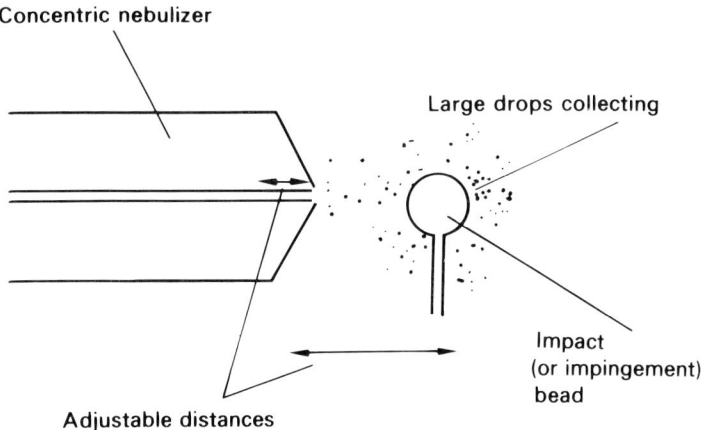

Fig. 2.6 Impact bead assembly.

gases. The chamber also allows for mixing of the gases and tends to damp fluctuations in nebulization efficiency. Some loss of solvent by **evaporation** will also occur. The chamber requires a **drain** tube which must be sealed to provide a back pressure for the flame. This is usually achieved by a hydraulic seal (a 'U' tube). It is imperative that water is present in the 'U' tube before lighting the flame!

The **efficiency** of this kind of indirect nebulizer and expansion chamber arrangement is usually about 10%. It is favoured because only very small droplets reach the flame, but clearly it would be nice to improve its efficiency. The use of **organic solvents** is known to do this. Organic solvents may affect nebulization because of their different surface tension, density, viscosity or saturated vapour pressure, all of which enter into the relevant equations. They also affect results because of their effect on the flame (they essentially act as secondary fuel). Therefore, it is not surprising that there has been much controversy in the literature about these effects. L'vov has dealt with this subject well (see Appendix C). It seems likely that the several-fold increase in nebulization efficiency when using organic solvents is largely due to smaller droplets being formed (the surface tension being lower) which hence evaporate more readily.

Ultrasonic nebulizers have been used for flame work, but they are complex and have a tendency to 'memory' effects, i.e. traces of previous solutions contaminate successive samples. These are, therefore, not presently used commercially.

At this stage, no atoms have been formed—only a mist. Hence the name nebulization, not atomization. Some manufacturers use the term atomizer for nebulizer. This is more appropriate to the less rigorous perfumery industry. We will now discuss **atomization**.

The **desolvation** of the droplets is usually completed in the pre-heating zone. The mist of salt clotlets then fuses and **evaporates** or sublimes. This is critically dependent on the size and number of the particles, their composition and the flame mixture. In section 6.2, we shall discuss problems which can arise in analytical situations. As the absolute concentration of analyte in the flame is very small ($<10^{-3}$ atm), the saturated vapour pressure may not be exceeded even at temperatures below the melting point.

Many of these vapours will **break down** spontaneously to atoms in the flame. Others, particularly diatomic species such as metal **monoxides** (e.g. alkaline-earth and rare-earth oxides), are more refractory. Monohydroxides which can form in the flame may also give problems. The high temperature and enthalpy of the flame aids **dissociation** thermodynamically, as does a reducing environment. The role of **flame chemistry** is also important. Atoms, both ground state and excited, may be produced by radical reactions in the primary reaction zone. During recent work on determining cerium in steels, we showed the importance of this in atomizing and exciting the very refractory rare-earth oxides. If we take the simplest flame (a hydrogen–oxygen flame), some possible reactions are:

FLAME SPECTROSCOPY

$$H_2 + Q \rightarrow 2H^\bullet$$
$$O_2 + Q \rightarrow 2O^\bullet$$
$$H^\bullet + O_2 \rightarrow O^\bullet + HO^\bullet$$
$$O^\bullet + H_2 \rightarrow H^\bullet + HO^\bullet$$
$$H^\bullet + H_2O \rightarrow H_2 + HO^\bullet$$
$$(Q = \text{energy})$$

So far, this is most unflame-like, because energy has only been consumed! Equilibrium is approached through third body (B) collisions such as:

$$H^\bullet + H^\bullet + B \rightarrow H_2 + B + Q$$
$$H^\bullet + {}^\bullet OH + B \rightarrow H_2O + B + Q$$

where B may be N_2, O_2 or analyte-containing molecules, e.g. NaCl:

$$H^\bullet + NaCl \rightarrow Na^\bullet + HCl$$
$$H^\bullet + HO^\bullet + NaCl \rightarrow H_2O + Na^\bullet + Cl^\bullet + Q$$
$$H^\bullet + HO^\bullet + NaCl \rightarrow H_2O + Na^{\bullet *} + Cl^\bullet$$

(* denotes an atom excited by 'chemiluminescence')

For many elements, the **atomization efficiency** (the ratio of the number of atoms to the total number of analyte species, atoms, ions and molecules in the flame) is 1, but for others it is less than 1, even in the nitrous oxide–acetylene flame (for example, it is very low for the lanthanoids). Even when atoms have been formed, they may be lost by compound formation or **ionization**. The latter is a particular problem for elements on the left of the periodic table (e.g. Na $\rightarrow$ Na$^+$ + e$^-$; the ion has a noble gas configuration, is very difficult to excite and so is lost analytically). Ionization increases exponentially with **temperature**, such that it must be considered a problem for the alkali, alkaline-earth and rare-earth elements as well as some others (e.g. Al, Ga, In, Sc, Ti, Tl) in the nitrous oxide–acetylene flame. Fortunately, this loss of atoms can be simply overcome by increasing the partial pressure of electrons in the flame. Thus, we observe some **self-suppression** of ionization at higher concentrations. For trace analysis, an **ionization suppressor** or **buffer** consisting of a large excess of an easily ionized element (e.g. caesium or potassium) is added. The excess caesium ionizes in the flame, suppressing ionization (e.g. of sodium) by a simple, mass action effect:

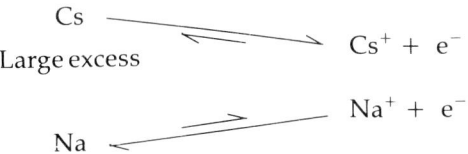

Differing amounts of easily ionizable elements in real samples cause varying ionization suppression and hence the possibility of interference (see section 6.4.2).

AN INTRODUCTION TO ATOMIC ABSORPTION SPECTROSCOPY

Q. How does the venturi effect contribute to the nebulization of the sample?

Q. In the type of nebulizer described, where does most of the sample go?

Q. How can ionization problems be overcome?

Q. Summarize nebulization and atomization in a schematic diagram. Now compare this with Fig. 2.7.

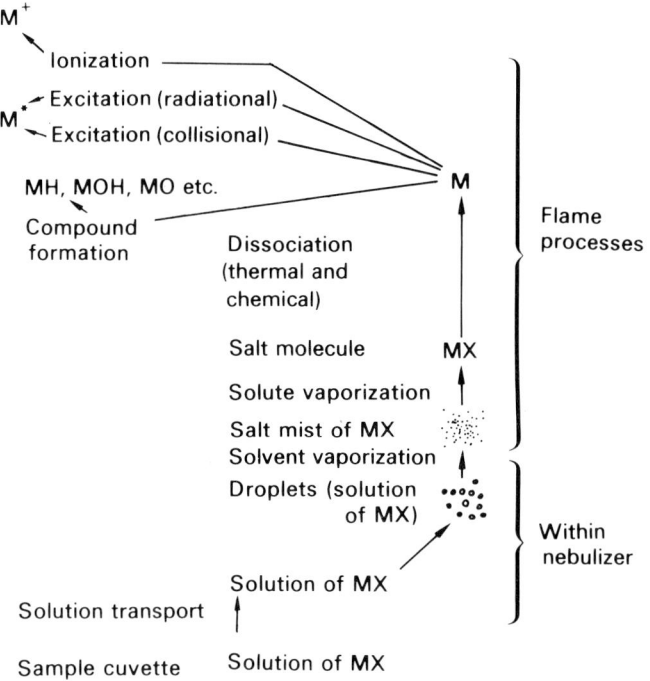

Fig. 2.7 Summary of atomization in flames.

2.5 BURNER DESIGN

Many early experiments were carried out with the **Meker type** of burner. A perforated metal plate is placed on the top of an open pipe. (The thicker the plate, the more stable the flame.) On ignition, a number of small cones are formed over each opening. During design, the number of holes, their size and the conductivity of the metal must be considered. Such burners are best suited to flames of low burning velocity, but are still occasionally used for atomic emission and atomic fluorescence spectroscopy.

Slot burners are now far more popular. Slots of 100 mm are popular for atomic absorption spectroscopy using air–acetylene but, because of the higher burning velocity of nitrous oxide–acetylene, such a long slot is not safe and one of 50 mm is used. The width of the slot, its length and the conductivity of the metal used (commonly aluminium, stainless steel or titanium) are important. The narrower the slot, the greater the cooling and stability. However, the tendency to clog is increased. **Three-slot** (or Boling) burners, with three parallel slots, are also available. A profile of a well designed, single-slot burner is shown in Fig. 2.8. The raised edges at the slot help to prevent carbon build-up; the curvature at the base helps to avoid turbulence.

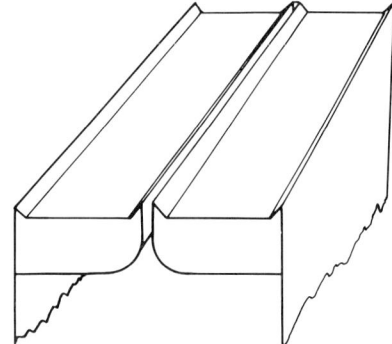

Fig. 2.8 Cutaway diagram to show the profile of a single-slot burner with raised edges.

It is now recognized that slot burners are also advantageous for atomic emission spectroscopy at the trace level (where self-absorption (see section 3.1) is not a problem) for viewing reasons. **Circular** burners are preferred for atomic fluorescence and circular slot burners are available.

Q. Why is the slot used for a nitrous oxide–acetylene burner shorter than that used for an air–acetylene burner?

Q. How is it possible to redesign a slot burner to take up to 40% solids (cf. the normally acceptable 4–6%) when using a highly conductive material for the burner head?

2.6 FLAME SPECTRA

Figure 2.9 shows the **spectrum of the air–acetylene** flame. The strong OH bands and the C_2 'Swann bands' are significant. Most of these molecular species are associated with the secondary zone. In the nitrous oxide–acetylene flame, the CN spectrum is also seen clearly.

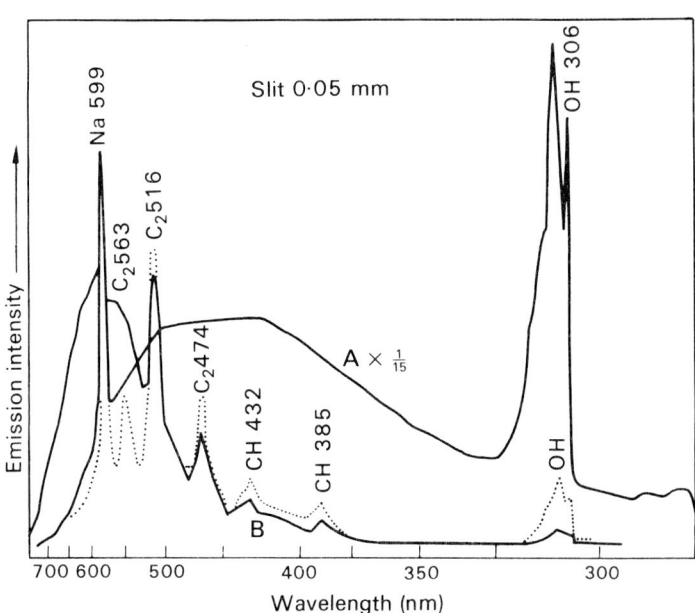

Fig. 2.9 Background emission from unseparated (A) and separated (B) air–acetylene flame. The ordinate of A should be multiplied by 15 to compare absolute intensities. The effect of decreased fuel flow rate on emission from separated flame is shown by broken line.

The secondary zone can be **separated** from the inter-conal zone by preventing the diffusion of air into the flame. Figure 2.10 shows two ways in which this can be done, either mechanically, using a silica tube or, for better analytical performance, by a laminar shield of nitrogen or argon flowing round the flame. Argon-sheathed **separated flames** have particular use in atomic fluorescence spectroscopy, where it is advantageous to measure small signals against a low background and also reduce quenching.

FLAME SPECTROSCOPY

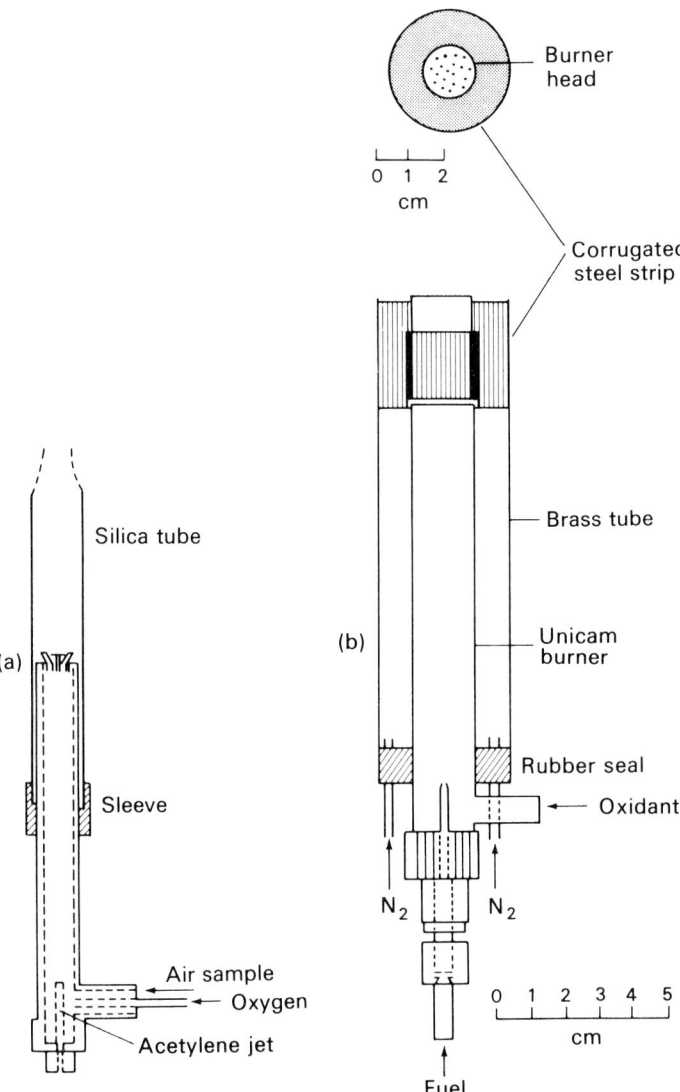

Fig. 2.10 (a) Burner arrangement for separated air–acetylene flame employing silica separator tube. (b) Burner arrangement for separated air–acetylene flame employing nitrogen shielding gas.

Q. Why would you expect problems in measuring atomic fluorescence spectroscopy in the region 304–320 nm in an unseparated flame?

2.7 BROADENING

Now we have the atoms in the flame, we must consider one further detail. Atomic lines are not infinitely thin and their width is best discussed by talking about the 'half-width' ($\Delta \nu$ cm^{-1}). The half-width is defined in Fig. 2.11. L'vov gives an excellent treatment of this subject (see Appendix C).

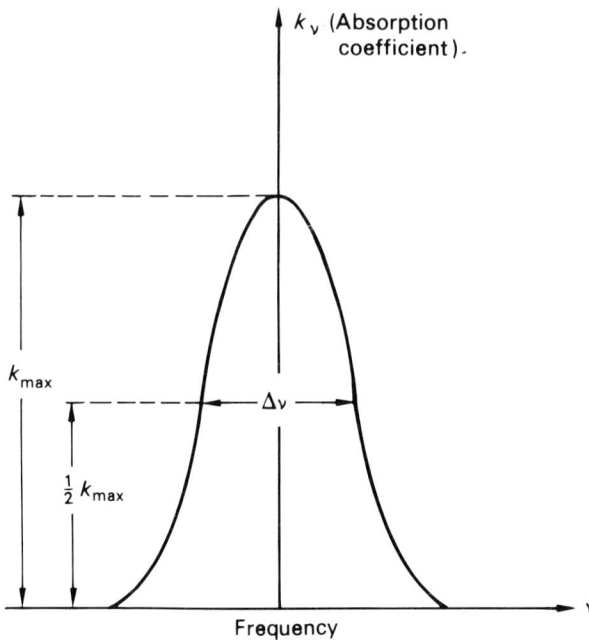

Fig. 2.11 Profile of an absorption line. The half-width $\Delta \nu$ is the width of the line when $k_\nu = \tfrac{1}{2}k_{max}$.

Natural broadening comes about because of the finite lifetime (τ) of the atom in the excited state. By **Heisenberg's Uncertainty Principle**, if we know the state of the atom, we must have uncertainty in the energy level.[1] This type of broadening is negligible in flames.

Doppler broadening arises from the random thermal motion of the atoms relative to the observer. The velocity V_x of an atom in the line of sight will vary according to the **Maxwell distribution**, the atoms moving in all directions relative to the observer.[2]

---

[1] We assume τ for the ground state is infinity and, therefore, for a **resonance line** (a line arising from a transition where the ground state is the lower state), the natural width, $\Delta \nu_N = 1/2\pi\tau$.

[2] The frequency will be displaced by $\Delta \nu = (V_x/c)\nu_0$. To this is applied the distribution of velocities. After evaluation of constants, this simplifies to the Doppler half-width, $\Delta \nu_D = 7.16 \times 10^{-6}\nu_0\sqrt{(T/M)}$ where M = relative atomic mass.

Lorentz or collisional broadening arises from collision of atoms with atoms, or with molecules, of a different kind. Experimental data show that these collisions shift, broaden and cause asymmetry in the line. Different gases have different effects. Collisional theory offers the best fit equations to describe these events at the line centre, and statistical theory describes the events at the wings. Lorentz broadening increases with pressure (P) and temperature (T), and is generally regarded as being proportional to P and $\sqrt{T}$ (or $T^{3/10}$, in some treatments).

The **absorption profiles** of analyte atoms in flames are generally considered to be governed by Doppler and Lorentz broadening, the centre primarily by the Doppler effect and the wings by the Lorentz effect.[3] Thus, $\Delta\nu$ **increases with T and P**.

To complete the story, we should mention that there are other broadening processes. **Holtsmark or resonance broadening** arises from collisions between atoms of the same kind and is therefore of little importance compared with other collisions. **Stark broadening** occurs in the presence of electric fields and is not applicable to flames; nor, usually, is **Zeeman broadening** (but see section 4.2.3), which occurs in magnetic fields.

Hyperfine structure (h.f.s.), due to the interactions of the spin of the nucleus, I, and the resultant spin moment of the electrons, J, or to the presence of several **isotopes**, means that we cannot talk of single lines, but of several very close, probably overlapping, 'lines'.[4]

For resonance lines, **self-absorption broadening** may be very important, because it is applied to the sum of all the factors above. As the maximum absorption occurs at the centre of the line (see Fig. 2.11), proportionally more

---

[3] The profile of lines can often be summarized by the Voigt profile:

$$k_\nu = k_0^{(D)} \frac{a}{\pi} \int_{-\infty}^{\infty} \frac{\exp(-y^2)\,dy}{a^2 + (w-y)^2}$$

where

$$a = \frac{\Delta\nu_L}{\Delta\nu_D}\sqrt{\ln 2} \qquad w = \frac{2(\nu - \nu_0)}{\Delta\nu_D}\sqrt{\ln 2}$$

$$y = \frac{2(\nu - \nu_0)}{\Delta\nu_D}\sqrt{\ln 2}$$

where $\Delta\nu_L$ is the Lorentz half-width.

[4] For example, the Hg 253.7 nm line consists of ten components (five even-numbered isotopes; ^{199}Hg, $I = \frac{1}{2} \to 2$ components; ^{201}Hg, $I = \frac{3}{2} \to 3$ components), some of which coincide, leaving an h.f.s. of five observable components. Generally, we apply the following rules:

if	$\Delta\nu_{hfs} \ll \sqrt{(\Delta\nu_L^2 + \Delta\nu_D^2)}$	$\Delta\nu_{hfs}$ ignored
if	$\Delta\nu_{hfs} \gg \sqrt{(\Delta\nu_L^2 + \Delta\nu_D^2)}$	treat each line as simple and separate
if	$\Delta\nu_{hfs} \approx \sqrt{(\Delta\nu_L^2 + \Delta\nu_D^2)}$	sum the individual profiles graphically

intensity is lost on self-absorption here than at the wings. Thus, as the concentration of atoms in the atom cell increases, not only the intensity of the line, but also its profile, changes, as shown in Fig. 2.12. High levels of self-absorption can actually result in **self-reversal**, i.e. a minimum at the centre of the line. This can be very significant for emission lines and spectral light sources.

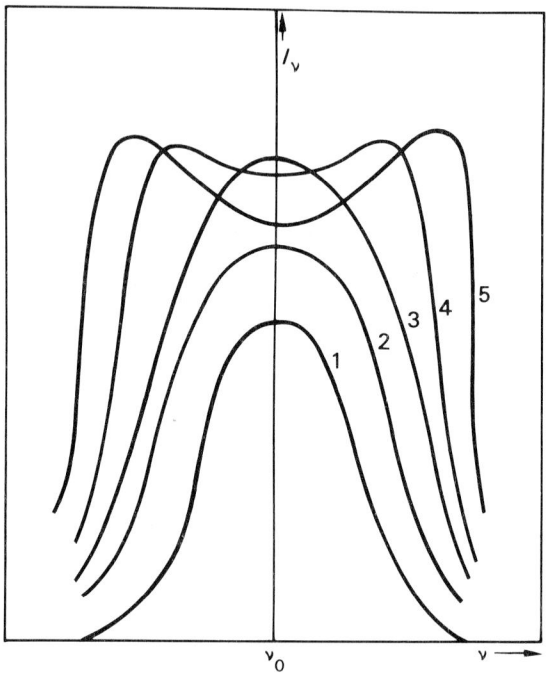

Fig. 2.12 Effect of self-absorption on line profile as concentration increases from 1 to 5.

Q. What two broadening processes have most control over absorption profiles in flames?

Q. What happens to the line profile as (i) the temperature increases; (ii) the pressure in the atom cell increases?

Q. What important additional factor may control the profile of an emission line?

3 FLAME EMISSION SPECTROMETRY

Usually, it is the radiation emitted by **atoms** which is observed, but that from other species is sometimes measured, hence the more general title above. Recently we have had considerable success in our laboratories in determining rare-earth elements in steels, using the emission at **oxide bands** as well as atomic emission. In cooler flames, sulphur may be determined—for example, using the blue S_2 emission. Generally, however, the notes below are concerned with atomic emission spectroscopy.

3.1 THEORY

The **probability** of transitions from given energy levels of a fixed atomic population was expressed by **Einstein** in the form of three **coefficients**. These are termed **transition probabilities**, A_{ji}, B_{ij}, B_{ji}, which refer to spontaneous emission, absorption and stimulated emission, respectively. They can be considered as representing the ratio of the number of atoms undergoing a transition to the number in the initial level. The intensity I_{em} of a **spontaneous emission** line is related to A_{ji} by the equation:

$$I_{em} = A_{ji} h \nu_{ji} N_j \tag{3.1}$$

When a system is in thermodynamic equilibrium, the level population, i.e. the number of atoms N_j in the excited state, is given by the **Boltzmann Distribution Law**:

$$N_j = \frac{N_o g_j}{g_o} \exp[-(E_j/kT)] \tag{3.2}$$

where N_o is the number of atoms in the ground (unexcited) state with an energy $E_o = 0$, and g_j and g_o are the statistical weights of the jth and ground states, respectively, where $(g = 2J + 1)$, J is the third quantum number and k is a constant. Thus:

$$\frac{N_j}{N_o} = \frac{g_j}{g_o} \frac{\exp[-(E_j/kT)]}{\exp[-(E_o/kT)]} \qquad (3.3)$$

If we express N, the total number of atoms present, as the sum of the population of all levels, i.e. $N = \Sigma_j N_j$:

$$\frac{N_j}{N} = \frac{g_j \exp[-(E_j/kT)]}{\Sigma_j g_j \exp[-(E_j/kT)]}$$

$$= \frac{g_j \exp[-(E_j/kT)]}{F(T)} \qquad (3.4)$$

where $F(T)$ is known as the **partition function**.

If self-absorption is neglected for a system in thermodynamic equilibrium:

$$I_{em} = A_{ji} h\nu_{ji} \frac{N g_j \exp[-(E_j/kT)]}{F(T)} \qquad (3.5)$$

A similar result is more readily, if less rigorously, obtained if we assume that virtually all the atoms remain in the ground state (the strength of this assumption can be seen in Table 3.1). Thus, N, the total number of atoms which is directly related to the concentration in solution, is approximately equal to N_o. Thus, Eqn (3.2) becomes

$$N_j = N \frac{g_j}{g_o} \exp[-(E_j/kT)] \qquad (3.6)$$

and Eqn (3.1) becomes:

$$I_{em} = A_{ji} h\nu_{ji} N \frac{g_j}{g_o} \exp[-(E_j/kT)] \qquad (3.7)$$

This is similar to Eqn (3.5) for practical purposes and the reader may prefer this simplified derivation.

Thus, the intensity of atomic emission is critically dependent on **temperature**. It also follows that when low concentrations of analyte atoms are used (i.e. when self-absorption is negligible), the plot of emission intensity against sample concentration is a **straight line**. (A fuller treatment of the theory is given by Sharp (see Appendix C) in his review of plasma emission spectrometry.)

The effect of temperature and E_j, the energy difference between the excited and ground states, is perhaps best illustrated by the practical examples in Table 3.1.

It can clearly be seen that the number of excited atoms, and hence the intensity of emission, increases very rapidly with increasing **temperature**, and that the number of excited atoms is greater the lower the **energy level**. In spite of this, the number of excited atoms at typical flame temperatures is very low indeed compared with the number of ground state atoms, even for easily excited lines. For a difficult-to-excite line (e.g. Zn 213.9 nm), it can be

FLAME EMISSION SPECTROMETRY

Table 3.1

Element	Line (nm)	g_i/g_j	E_j (eV)	Ratio of excited to ground state atoms			
				2000 K	3000 K	4000 K	5000 K
Cs	852.1	2	1.46	4.44×10^{-4}	7.24×10^{-3}	2.98×10^{-2}	6.82×10^{-2}
Na	589.1	2	2.11	9.86×10^{-6}	5.88×10^{-4}	4.44×10^{-3}	1.51×10^{-2}
Ca	422.7	3	2.91	1.21×10^{-7}	3.69×10^{-5}	6.03×10^{-4}	3.33×10^{-3}
Zn	213.9	3	5.7	7.29×10^{-15}	5.58×10^{-10}	1.48×10^{-7}	4.32×10^{-6}

shown that only about one excited atom will exist at any given time in an air–propane flame when aspirating a 1 mg l^{-1} zinc solution.

Figure 3.1 shows a **Grotrian diagram**, or partial energy level diagram, for sodium emission. If the oscillator strengths of the lines were equal (which, of course, they are not), we would expect to see maximum emission from lines where the upper excited state lines lie closest to the ground state. That is, where the excitation energy E_j is small and the Boltzmann distribution predicts a greater population of excited atoms than at other states. The 3P–3S transition for sodium is very intense—the famous D lines—and hence it is an element very favourable for analysis by atomic emission spectrometry.

As the concentration of atoms in the flame increases, the possibility increases that photons emitted by excited atoms in the hot region in the centre will collide with atoms in the cooler, outer regions of the flame, and thus be absorbed. This **self-absorption** effect contributes to the characteristic **curvature** of atomic emission calibration curves towards the concentration axis, as illustrated in Fig. 3.2. For these reasons, flame atomic emission spectrometry is only suitable for trace analysis, unless non-resonance lines are used.

Q. Why is the greatest analytical sensitivity observed in flame AES with (i) alkali metals; (ii) in a nitrous oxide–acetylene flame?

Q. Why do flame AES calibration curves typically show more curvature at equivalent concentrations, but extend to lower concentrations when using a slot burner, viewed end-on, compared with a small circular burner?

3.2 INSTRUMENTATION

A **wide range** of instrumentation may be employed for atomic emission spectroscopy, from filter photometers to highly sophisticated, custom-built instruments.

Flame photometers are frequently used for alkali metal determinations in

AN INTRODUCTION TO ATOMIC ABSORPTION SPECTROSCOPY

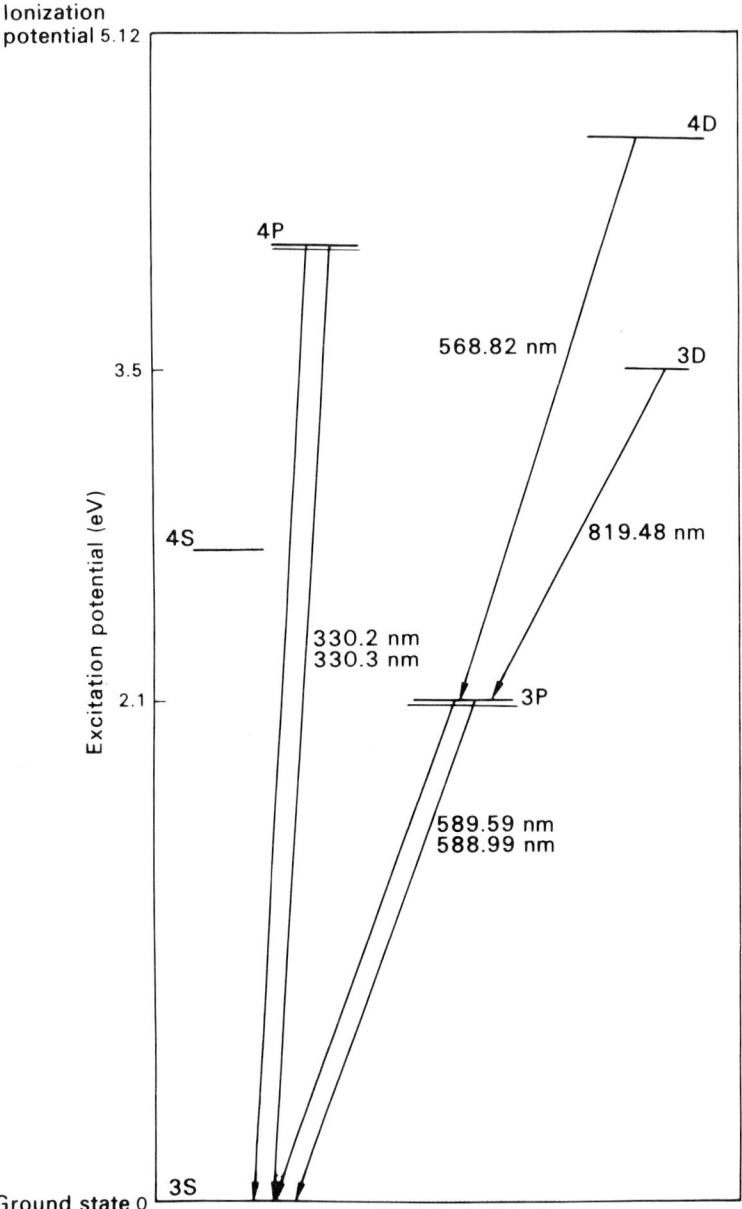

Fig. 3.1 Simplified Grotrian diagram for sodium (emission).

clinical and agricultural analyses. A low temperature (e.g. air–natural gas) flame is used; thus only the most prominent lines are excited. These lines are isolated by coloured glass **filters** (usually labelled K, Li, Na etc.). A **bar-**

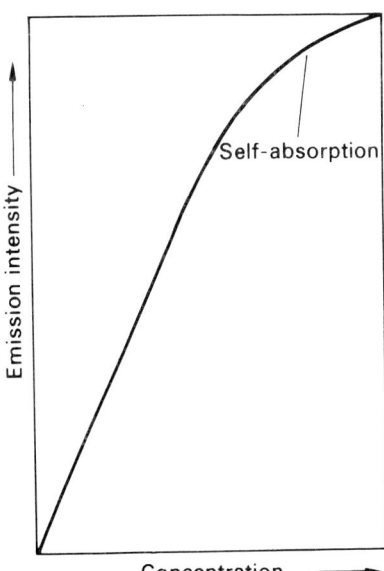

Fig. 3.2 Characteristic flame atomic emission calibration curve.

rier layer photocell[1] may be used as the detector, with read-out on a galvanometer.

More sophistication may be introduced by using **interference filters** instead of colour filters, and **vacuum phototubes**, or **photomultipliers**, as detectors. In one instrument, lithium is added as an internal standard to the sample, usually clinical and hence lithium free. The emitted light beam is split, one half passing through a 'lithium filter' and the other half through a filter which transmits only the analyte line. The power of the two beams is then ratioed, thus allowing for variations in parameters such as uptake rate and flame temperature.

Much atomic emission work is performed on atomic absorption instruments, and the author has shown such instruments to be quite suitable for the analysis of trace metals in steel. A nitrous oxide–acetylene flame is used and the instrument switched to the **emission mode**. Certain solution spectrophotometers have **flame attachments** for atomic emission work.

High resolution is advantageous in atomic emission spectroscopy (see section 6.2.1), and the typical ½ or ¼ metre grating of an atomic absorption instrument does not offer this. Therefore, **echelle grating** monochromators (see section 1.3) have been used for atomic emission spectroscopy.

---

[1] A plate on which a thin layer of a semiconductor (e.g. selenium) has been deposited. A very thin transparent layer of silver is sputtered over the selenium to act as a collector electrode. Light falling on the semiconductor surface excites electrons which are released to the collector electrode. The current thus generated is measured using a galvanometer.

AN INTRODUCTION TO ATOMIC ABSORPTION SPECTROSCOPY

Flame background and **molecular emission** pose particular problems in atomic emission spectroscopy, especially if organic solvents are used. The simplest **correction** which can be made is to subtract the background, blank reading from the reading obtained when the analyte solution is sprayed into the flame. As the analysis solution may alter the background, it may be preferable to measure the background at a second, nearby wavelength whilst spraying the analyte solution. The background may, however, be non-linear, e.g. the analysis line may be adjacent to a molecular band. Corrections might then be better applied by measuring the background an equal distance either side of the analysis line and subtracting the mean of these two readings. A **wavelength scan** attachment greatly aids this correction, as the line can be scanned a few nanometres either side, automatically and reproducibly. These options are summarized in Fig. 3.3.

If the wavelength is scanned repetitively and rapidly, backwards and forwards, a **derivative** signal is obtained. This can be achieved either by vibrating the slits or by placing a vibrating (or stepped) **refractor plate**, usually made of quartz, in the light path. The background, being roughly constant over a few nanometres, gives rise to a very small derivative signal, but the line, being a sharp maximum, gives rise to a large signal (see Fig. 3.4). If the amplifier is **tuned** to a frequency of $2f$, where f is the vibration frequency, the signal from the line can be selectively amplified. This technique is known as the **wavelength modulation** technique and an early account is given by Snelleman et al. (*Anal. Chem.* **42**, 394 (1970)).

Q. Why are (i) internal standards; (ii) higher resolution monochromators; (iii) high temperature flames; (iv) background correction, used in AES?

3.3 OTHER SOURCES FOR ATOMIC EMISSION SPECTROSCOPY

3.3.1 Introduction

We cannot leave the subject of atomic emission spectroscopy without mentioning some of the other excitation sources frequently used. Some of these sources are more widely employed for atomic emission than flames, and constitute an area worthy of a similar book in their own right. Therefore, their treatment here must be necessarily brief. Many of these sources are used normally for **simultaneous multi-element** measurements. In this mode, they are no longer used with monochromators, but with **polychromators**, i.e. a prism or grating dispersion apparatus with several exit

FLAME EMISSION SPECTROMETRY

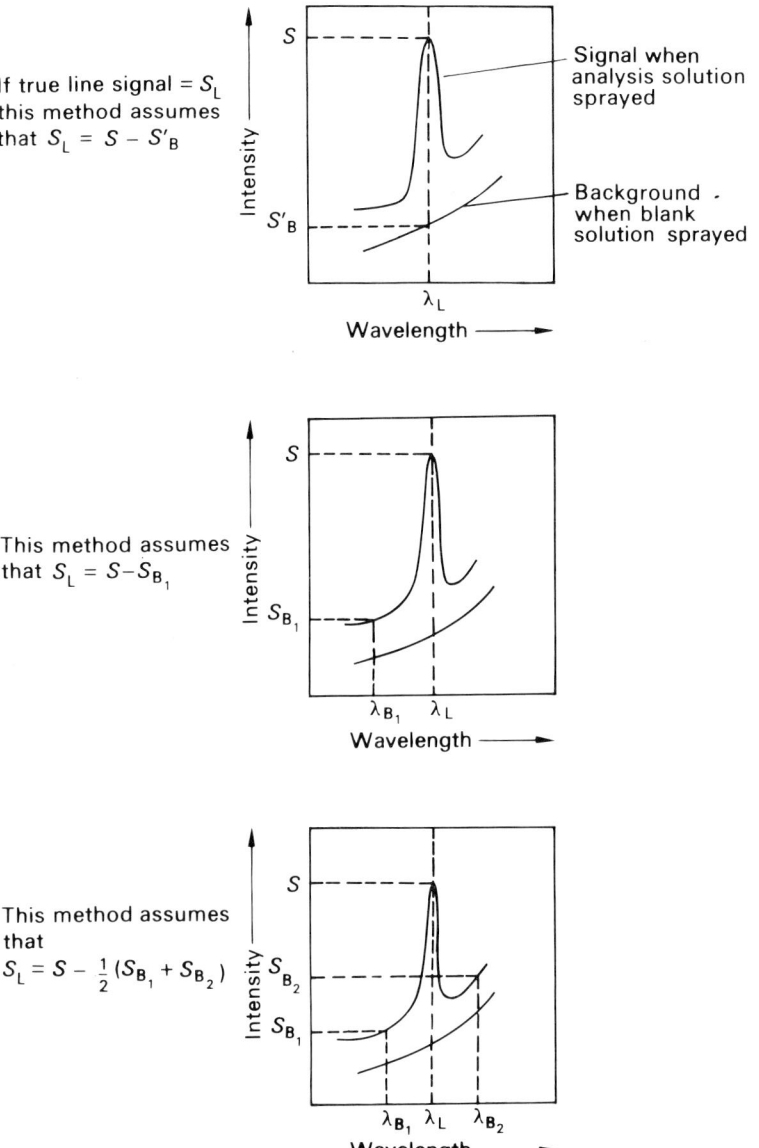

Fig. 3.3 Background correction methods used in atomic emission spectroscopy.

slits, usually fixed at the correct angle for a pre-determined set of elements (commonly 20–30).

The first group of techniques is more suited to work with **solid samples**, and liquid samples can only be accommodated if special precautions are taken. The second group consists of what are generally regarded as solution

AN INTRODUCTION TO ATOMIC ABSORPTION SPECTROSCOPY

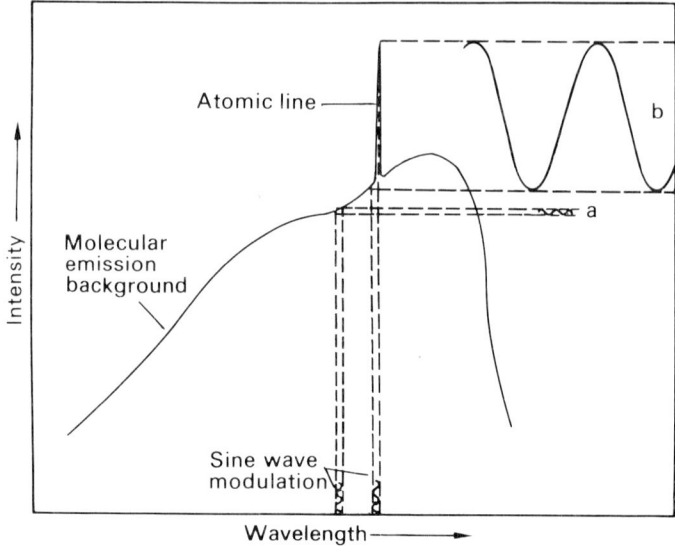

Fig. 3.4 Schematic representation of wavelength modulation. (a) Molecular emission having a band structure gives rise to a small AC component. (b) The sharp peak of the atomic line gives rise to a large AC component.

techniques, which require modification for work with solid samples. The use of solid samples greatly speeds **sample preparation** (e.g. by allowing spark analysis of a complete steel sample within 90 s of it being taken from a furnace) and avoids the problems of dilution and contamination inherent in dissolution. However, solid sampling usually presents more problems in avoiding errors from surface contamination and unrepresentative or inhomogeneous samples, and in assuring complete or representative atomization. Flames are not commonly used for solid samples and are therefore not seen as rivals to the first group of techniques. In any case, many samples are liquids.

Q. When is a polychromator used instead of a monochromator?

Q. What are the advantages and disadvantages of using solid samples?

3.3.2 Solid sampling emission sources

An arc is a continuous **electrical discharge** of high energy between two electrodes. In the **DC arc**, the sample is usually packed into an **anode** electrode, or the sample itself becomes the anode, as, for example, in metals analysis.

The arc is struck and the sample is **vaporized** into the discharge region, where the excitation and emission occur. Because relatively large amounts of sample are excited, the obtainable detection limits are low, but the **unstable** nature of the discharge leads to poor precision. Additionally, the intensity of emission is highly dependent upon the **matrix**. If the light from the discharge is dispersed by a prism, and a camera of long focal length (e.g. 1 m) is used, a **photograph** of the spectrum gives good resolution and can be used for the rapid qualitative identification of the elements in an unknown sample. The technique, **spectrography**, may be made more quantitative, but nowadays the DC arc is most commonly used qualitatively, with the possible exception of in geological analysis.

If an **intermittent** discharge is used (e.g. an **AC spark**) the precision can be greatly improved. However, the detectability is poorer because only relatively small amounts of sample will be vaporized. Modern electrical discharges combine the characteristics of both the arc and spark (e.g. a unidirectional spark may be used so that the intermittent discharge always excites the sample as the cathode) to obtain optimal detectability and precision for quantitative work. Further improvements have been made by **sheathing** the electrodes with argon to reduce the amount of self-absorption by cooled sample, reduce background and stabilize the discharge. Attempts to reduce the influence of surface effects on the final result by careful pre-sparking are also made. Such discharges are used very widely in large **polychromator** systems, with **computer control** of the data acquisition and calculation of the several corrections needed because of inter-element effects and spectral **interferences**. These highly expensive instruments are referred to as **direct reading spectrometers** and are now indispensible to the operation of, for example, most modern integrated steelworks.

Growing interest is being shown in emission sources where the sample is part of a lamp. In the **Grimm discharge lamp**, or **glow discharge lamp**, the sample is clamped on to the end of the discharge chamber as a **conducting solid disc** (non-conducting samples must be mixed with a conducting material and pressed into a pellet). This is then flushed with inert gas and **evacuated** to a pressure of 500 Pa. The discharge is initiated and observed through a quartz window opposite the sample. Extended linear calibrations, **excellent precision** and minimal interferences are obtained. Work is proceeding to try to improve the detection limits, which are already competitive with flame emission.

A **hollow cathode lamp** (see also section 4.2.1) may be used as an emission source if the cathode is machined from the sample or small pieces are placed in the cathode. Again, the cathode is placed in a **low pressure** chamber which has been flushed with inert gas. When current passes in the lamp, sample **sputters** off the walls of the cathode and may be further excited by collisions. The resulting emission is observed through a quartz window. Although not offering the same excellent precision as the glow discharge lamp, the hollow cathode lamp does have good analytical characteristics and the special advantage of detection limits for **volatile elements** (such as arsenic, antimony and lead) orders of magnitude superior to those obtainable by flame emission spectroscopy. Both these lamps can be operated in conjunction with a direct reading spectrometer, but both suffer from the disadvantage of the need for the source to be **dismantled** between successive samples.

Q. What is the major contemporary role of the DC arc?

Q. What is a direct-reading spectrometer?

Q. What are the advantages of (i) the glow discharge lamp; (ii) the hollow cathode lamp? From what disadvantage do they both suffer?

3.3.3 Plasma sources

The four sources considered below are all referred to as **plasmas**. A plasma is defined as 'any **luminous** volume of gas having a fraction of its atoms or molecules **ionized**'. Although this definition encompasses flames, it is usually applied to plasmas formed by electrical excitation. Non-flame-like plasmas are confined to a **column** joining the current-carrying electrodes, e.g. the DC arc and AC spark discussed above. **Flame-like plasmas** exist with a significant part of the discharge **transferred from** the core into which the power is coupled. Thus, they are generally larger, and this enables **spatial resolution** of the plasma discharge and the atomic emission arising from the sample, ease of sample introduction and an ability to accept **liquid samples**.

The **DC arc plasma** is a **transferred plasma**, where a portion of the discharge has been moved out of the primary arc column. This can be achieved by altering the straight-line geometry of the electrodes and by the careful use of **gas flows** to bend the arc. Historically, a popular design had been to place the electrodes at right angles and to fire a strong gas flow at the centre of the plasma. This flow contained the sample aerosol and produced an inverted 'V'-shaped plasma. Such a plasma tended to be unstable and move about. Figure 3.5 shows a popular commercial design of DC plasma where a **third electrode** has been added to **stabilize** the discharge.

Often referred to as an **inverted 'Y' discharge**, this plasma accepts liquid samples in aqueous or organic solvents. It is unlikely that the sample actually penetrates the highest temperature part of the discharge (7000–9000 K). In any case, the high **plasma continuum** does not permit observation in this region, which is made instead in the angle of the 'Y'. The **excitation temperature** of the sample is about **5500 K**. The DC plasma is comparatively **economical** to run, consuming less than 1000 W and about 8 l min^{-1} of argon. Unfortunately, there may be a **tungsten** background spectrum from

FLAME EMISSION SPECTROMETRY

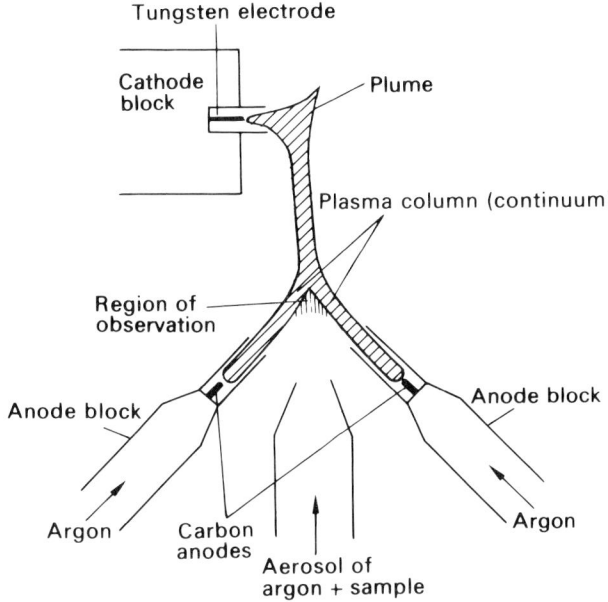

Fig. 3.5 DC Plasma (inverted Y plasma-jet).

the electrodes. A very troublesome problem with this plasma appears to be the substantial **enhancement** of emission caused by the presence of **easily ionizable** elements. Buffering with lithium or barium to overcome the problem has been reported, but with varying degrees of success. Easily ionizable elements may affect the plasma in a number of ways: (i) by reducing the plasma temperature by making it more conducting; (ii) by producing a more diffuse continuum; (iii) by emitting strongly itself; (iv) by shifting the ionization equilibrium towards atomization. The net effect seems to be a combination of these.

A popular commercial configuration uses the excellent resolution of an **echelle monochromator** (see section 1.3) (or **echelle polychromator**, if the exit slit is replaced by a cassette with about 20 elemental slits). The DC plasma offers precision and accuracy similar to flame emission and with improved detection limits for difficult-to-excite elements.

In the **capacitatively coupled microwave plasma**, a 2450 MHz magnetron valve is coupled, via a coaxial waveguide, to the plates or torch where the plasma is formed. Considerable problems have been encountered with this admittedly low-cost plasma, particularly from easily ionizable elements which cause dramatic changes in the excitation temperature of the plasma.

Alternatively, a **microwave-induced plasma** may be formed in a **resonant cavity** using a similar generator and powers up to 200 W. If a small flow of argon (e.g. 300 ml min^{-1}) is passed through a small bore (2 mm internal diameter) quartz tube in a microwave cavity and **seeded** with electrons from a high-voltage spark, a **self-sustaining** plasma will form. Several types of cavity have been used, including the ¼-wave **Broida** cavity and the ¼-wave **Evensen** cavity, but recently the TM_{010} **Beenakker** cavity has gained markedly in popularity because of its ability to sustain a helium plasma at atmospheric pressure. The **temperature** of this plasma is difficult to define as it is not in **local thermal equilibrium**, i.e. the various measured temperatures are not the same. The **excitation temperature** is in the region 5000–7000 K, but the **neutral gas temperature** is less than 1000 K. The presence, however, of high energy electrons and **metastable** excited inert gas species means that the microwave plasma is a highly efficient excitation source. The presence of highly excited helium metastables means that a helium plasma can even be used to produce line spectra for elements such as chlorine, fluorine, nitrogen and oxygen.

Although inexpensive and compact, the microwave-induced plasma suffers from a **low tolerance to solution** samples and **chemical interferences**. The latter arise both because of problems with easily ionized elements and because of the low gas temperature. The future of this plasma now seems to lie largely as a detector for **gas chromatography**, or in association with other devices, such as electrothermal atomizers, which provide a gaseous sample.

By far the most promising and logical extension to the chemical flame as an emission source is the **inductively coupled plasma** (ICP). Pioneered by Greenfield and Fassel, this plasma has clear advantages over other emission sources for solution work, provided that the **capital cost** and **running expenses** can be met. Figure 3.6 shows a schematic diagram of an ICP **torch**. The torch consists of three concentric tubes, the outer two usually being made of quartz. This is placed in the **work coil** of a **radiofrequency generator**. The outer gas flow is delivered **tangentially** and may be of either argon or nitrogen; in the latter case, it serves only to cool the outside of the plasma and protect the torch. The intermediate flow may be omitted if argon is used in the outer flow, as the outer gas can serve to propagate the plasma. Alternatively, an intermediate tangential flow of argon is used as the **plasma gas**. The central flow is known as the **injector gas** and this consists of argon plus the sample aerosol. The diameter of the torch is usually either 27 or 18 mm (the 'Greenfield torch' and the 'Fassel torch', respectively).

The radiofrequency generator may produce between 2 and 30 kW forward power at between 5 and 50 MHz, the usual combination being a **few kilowatts** at **27.12 MHz**. When the power is switched on to the 2- or 3-turn induction coil, an **AC magnetic field** is generated axially through the coil. When the argon gas flowing through the outer tubes of the torch is seeded with electrons using a spark, these electrons accelerate in the field. Rapidly, the electrons reach ionizing energies and collisions with the gas in the field produce further breakdown and an avalanche effect. This occurs almost instantaneously and the magnetic field causes the ions and electrons to flow in

FLAME EMISSION SPECTROMETRY

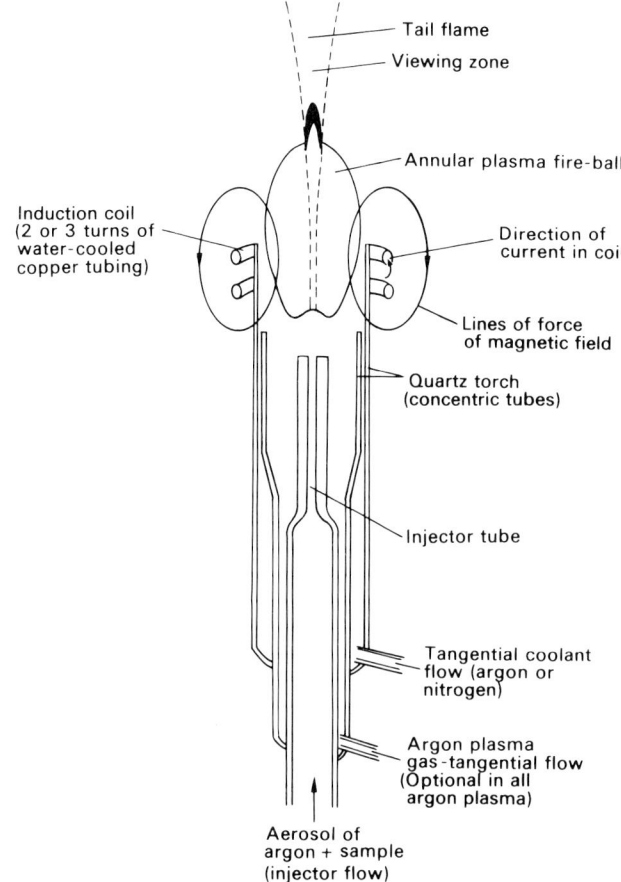

Fig. 3.6 Schematic diagram of an inductively coupled plasma torch.

closed circular horizontal paths. These **eddy currents** heat the neutral argon by collisional energy exchange and a white hot **fireball** is produced. The sample is introduced via the injector gas which punches a hole in the fireball, producing an **annulus** (commonly referred to as a doughnut).

The ICP is another plasma which is not in local thermal equilibrium—the gas temperature is less than the excitation temperature which is less than the ionization temperature. The fireball has been variously measured as being in the region 8000–10 000 K, but the strong continuum here means that better analytical signal-to-background ratio is obtained by measuring above this in the **tail-flame** where the temperature is 5000–6000 K.

Sample is normally introduced as an aerosol, following **ultrasonic** or **pneumatic nebulization** (see section 2.4). Ultrasonic nebulizers tend to give problems with memory effects and reliability. Therefore, despite the fact that only about 500 ml min^{-1} of argon nebulization gas is available, leading to nebulization efficiencies of only about 2%, pneumatic nebulization is commonly preferred. Specialized all-glass **miniature concentric** nebulizers (cf. flame nebulizers), **cross-flow nebulizers** (two capillaries at right angles, one carrying gas, the other sample) and **Babington-type** nebulizers (the sample drips down a groove containing a pin-hole through which the argon is introduced) are used. The aerosol passes through a cloud chamber, e.g. a double-pass chamber consisting of two concentric tubes. The nebulized gas is preheated by the ICP before it enters the plasma, where atomization and excitement occur. The way in which this happens is not yet fully understood, but it seems likely that the heat from the fireball (which acts like a tunnel oven) and collisions with highly excited metastable argon species play a role.

Initially, commercial plasma instruments combined an ICP with fixed wavelength **direct reading spectrometers**. While the extensive experience of such spectrometers ensures that they have reached a relatively trouble-free stage of development, they do lack versatility and require a high capital investment. More recently, instruments with computer-controlled, **rapid-scanning** monochromators have been marketed as an alternative way of exploiting, by sequential measurements, the multi-element advantages of the inductively coupled plasma.

The inductively coupled plasma offers several advantages in atomic spectroscopy. It will excite a **wide range** of metals and several non-metals, with **excellent limits of detection**. It will do this better than flame atomic absorption and flame emission, but not as well as electrothermal atomization, except for elements which are difficult to atomize, such as boron and uranium. The ICP enjoys excellent **freedom from interferences**, with an apparent absence of chemical interferences, controversy about the extent of ionization interferences, and equivalent problems with spectral interferences to those encountered with other emission techniques (i.e. good resolution is required). Long, linear **calibration ranges**, typically extending over 4 or 5 orders of magnitude, are obtained, and these are a decisive advantage for multi-element work, where different elements will be encountered at greatly varying concentrations. The ICP is more **reproducible** as an excitation source than is a flame, but **costs more** to run, consuming argon at 5–20 l min^{-1} and requiring a higher initial capital investment.

Clearly, the inductively coupled plasma has the advantage of a **higher excitation temperature** than a flame, and must be expected to yield lower detection limits than flame atomic emission, by consideration of the T term in Eq. (3.7), section 3.1. It can be seen that the ICP has other advantages and clearly will be a powerful, if more complex, future competitor with the techniques described in this book.

FLAME EMISSION SPECTROMETRY

Q. Define a flame-like plasma.

Q. How can a portion of a DC arc column be transferred to form a DC arc plasma?

Q. What are the advantages and disadvantages of the DC arc plasma?

Q. What interference problems may be encountered with microwave plasmas?

Q. Why are microwave-induced plasmas gaining in popularity as gas chromatographic detectors?

Q. How is an ICP formed?

Q. How may a sample be introduced into an ICP?

Q. Suggest ways in which atomization and excitation may occur in the ICP.

Q. Why are analytical measurements made in the tail-flame of the ICP?

Q. Why is the formation of an annular (doughnut-shaped) plasma considered essential for good analytical performance?

Q. What are the analytical advantages of the ICP?

4 FLAME ATOMIC ABSORPTION SPECTROSCOPY

Flame AAS (often abbreviated FAAS) is currently perhaps the most widely used method for trace metal analysis. It is particularly applicable where the sample is in solution or readily solubilized. It is very simple to use and, as we shall see, remarkably free from interferences. Its growth in popularity has been so rapid that on two occasions, the mid-1960s and the early 1970s, the growth in sales of atomic absorption instruments has exceeded that necessary to ensure that the whole face of the globe would be covered by atomic absorption instruments before the end of the century.

4.1 THEORY

Atomic absorption follows an exponential relationship between the intensity I of transmitted light and the absorption path length l, which is similar to Lambert's law in molecular spectroscopy:

$$I = I_o \exp(-k_\nu l) \qquad (4.1)$$

where I_o is the intensity of the incident light beam and k_ν is the absorption coefficient at the frequency ν. In quantitative spectroscopy, absorbance A is defined by:

$$A = \log(I_o/I) \qquad (4.2)$$

Thus, from Eqn (4.1), we obtain the linear relationship:

$$\begin{aligned} A &= k_\nu l \log e \\ &= 0.4343 k_\nu l \end{aligned} \qquad (4.3)$$

From classical dispersion theory (see Price (1979), p. 14, details in Appendix C), we can show that k_v is in practical terms proportional to the number of atoms per cubic centimetre in the flame, i.e. A is proportional to analyte concentration.

FLAME ATOMIC ABSORPTION SPECTROSCOPY

Atomic absorption corresponds to transitions from low to higher energy states. Therefore, the degree of absorption depends on the population of the lower level. When thermodynamic equilibrium prevails, the population of a given level is determined by Boltzmann's law (see section 3.1). As the population of the excited levels is generally very small compared with that of the ground state (that is, the lowest energy state peculiar to the atom), absorption is greatest in lines resulting from transitions from the ground state; these lines are called resonance lines.

Although the phenomenon of atomic absorption has been known since early last century, its **analytical potential** was not exploited until the mid-1950s. The reason for this is simple. Monochromators capable of isolating spectral regions narrower than 0.1 nm are excessively expensive, yet typical atomic absorption lines may often be narrower than 0.002 nm. Figure 4.1.1 illustrates this, but not to scale! The amount of radiation isolated by the conventional monochromator, and thus viewed by the detector, is not significantly reduced by the **very narrow atomic absorption signal**, even with high concentrations of analyte. Thus, the amount of atomic absorption seen using a **continuum source**, such as is used in molecular absorption spectroscopy, is negligible.

The contribution of Walsh was to replace the continuum source with an **atomic spectral source** (Fig. 4.1.2). In this case, the monochromator only has to isolate the line of interest from other lines in the lamp (mainly lamp filler-gas lines) (Fig. 4.1.3). In Fig. 4.1.4, we see that the atomic absorption signal exactly overlaps the atomic emission signal from the source and very large reductions in radiation are observed.

Of course, this exact **overlap** is no accident, as atomic absorption and atomic emission lines have the same wavelength. The very narrowness of atomic lines now becomes a positive advantage. The lines being so narrow, the chances of an accidental overlap of an atomic absorption line of one element with an atomic emission line of another is almost negligible. The uniqueness of overlaps in the Walsh method is often known as the **'lock and key' effect** and is responsible for the very high **selectivity** enjoyed by atomic absorption spectroscopy.

The best sensitivity is obtained in this method when the **source line is narrower** than the absorption profile of the atoms in the flame. Obviously, the other situation tends towards Fig. 4.1.1.

In recent years, it has been shown that the construction of atomic absorption spectrometers using **continuum sources** is possible, if somewhat expensive and complicated. So far, commercial manufacturers have not yet produced instruments of this type, which have remained the creations of a few research laboratories possessing very high resolution monochromators. Perhaps the most practical approach to continuum source atomic absorption has been by O'Haver and his colleagues (A. T. Zander *et al.*, *Anal. Chem.* **48**, 1166 (1976); J. M. Harnly *et al.*, *Anal. Chem.* **51**, 2007 (1979)). The basis of their system is a high-intensity (300 W) **xenon arc lamp**,

43

AN INTRODUCTION TO ATOMIC ABSORPTION SPECTROSCOPY

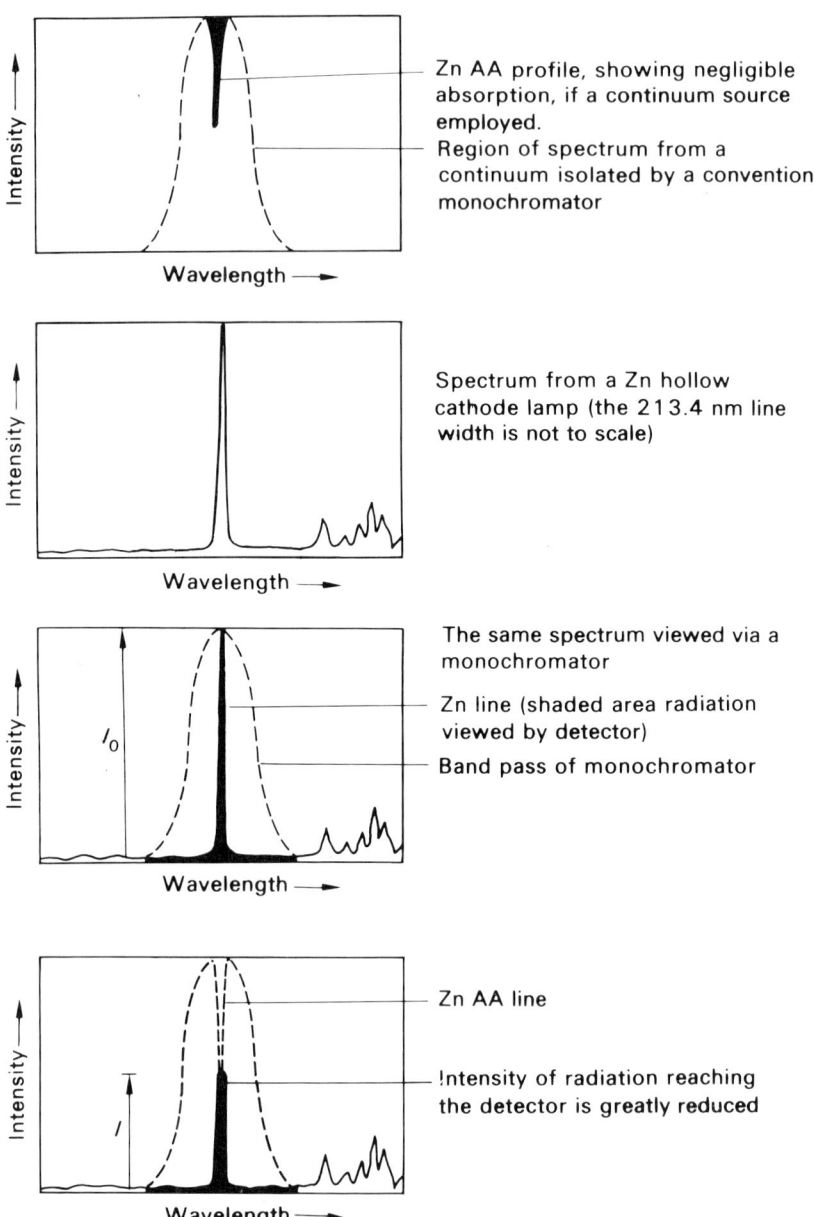

Fig. 4.1 Relative atomic absorption of light from continuum and line sources.

FLAME ATOMIC ABSORPTION SPECTROSCOPY

an **echelle grating** monochromator (see section 1.3) with **wavelength modulation** (see section 3.2), and an amplifier locked into the modulated signal. Competitive detection limits have been obtained by these workers, except for lines in the low ultraviolet, where the arc intensity is poor. The technique has the possibility of simple adaptation to **multi-element** work with in-built background correction.

A fuller account of atomic absorption theory is given by Kirkbright and Sargent (see Appendix C).

Q. Why is a plot of the percentage of light absorbed versus concentration a curve? What must be plotted to give a straight line passing through the origin?

Q. Why are resonance lines always used for analytical AAS?

Q. Why must a line source be used for AAS?

Q. How does the 'lock and key' effect impart great selectivity to AAS?

4.2 INSTRUMENTATION

Atomic absorption spectroscopy instrumentation can conveniently be considered under four headings: sources, atom cells, instrument design and read-out systems.

4.2.1 Sources

As we have seen, a **narrow line source** is required for AAS. Although in the early days vapour discharge lamps were used for some elements, these are rarely used now because they exhibit self-absorption. The most popular source is the hollow cathode lamp, although recently the use of electrodeless discharge lamps has grown in popularity.

The hollow cathode lamp

A hollow cathode lamp is shown diagrammatically in Fig. 4.2. As the name suggests, the central feature is a **hollow cylindrical cathode**, lined with the metal of interest. The lamp is contained within a glass envelope filled with an **inert gas** (usually Ne or Ar) at 1–5 torr. A potential of about 500 V is applied between the electrodes and, at the pressures used, the discharge concentrates into the hollow cathode. Typically, currents of 2–30 mA

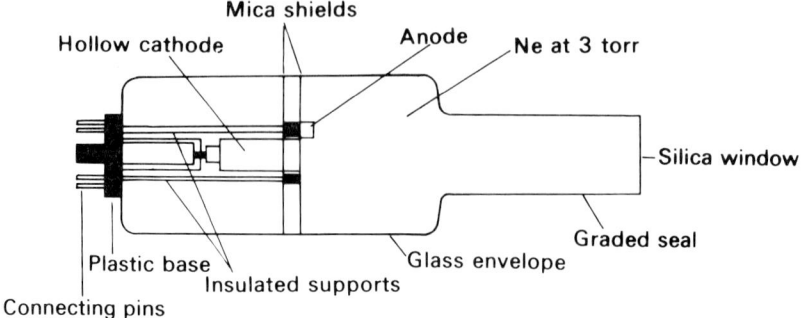

Fig. 4.2 The hollow cathode lamp.

are used. The filler gas becomes charged at the anode, and the ions produced are attracted to the cathode and accelerated by the field. The bombardment of these ions on the inner surface of the cathode causes metal atoms to **sputter** out of the cathode cup. Further collisions excite these metal atoms, and a simple, intense characteristic spectrum of the metal is produced. Christian and Feldman, and Kirkbright and Sargent (see Appendix C) describe this action and hollow cathode lamps in more detail.

The **insulation** helps to confine the discharge within the hollow cathode, thus reducing the possibility of self-absorption and the appearance of ion lines. Both these effects can cause bending of calibration curves towards the concentration axis. A glass envelope is preferred for ease of construction, but a silica window must be used for ultraviolet light transmission. A **graded seal** between the window and envelope ensures excellent gas-tightness and shelf life. A moulded plastic base is used. The choice of **filler gas** depends on whether the **emission lines** of the gas lie close to useful resonance lines and on the relative **ionization potentials** of the filler gas and cathode materials. The ionization potential of **neon** is higher than that of **argon**, and the neon spectrum is also less rich in lines. Therefore, neon is more commonly used.

Modern hollow cathode lamps only require a very short **warm-up** period. **Lifetimes** are measured in ampere hours (usually they are in excess of 5 A h). A starting voltage of 500 V is useful, but operating **voltages** are in the range 150–300 V. In many instruments, the current supplied to the lamp is **modulated**. Hollow cathode lamps may also be **pulsed** or run continuously. Hollow cathode lamps are comparatively free from self-absorption, if run at low current.

Normally, a different lamp is used for each element. **Multi-element lamps** (e.g. Ca–Mg or Fe–Ni–Cr) are available, but are less satisfactory due to the differing volatilities of the metals. **Demountable** (water-cooled) hollow cathode lamps are also marketed, but not widely used.

Electrodeless discharge lamps

Electrodeless discharge lamps were first developed for use in AFS. These lamps were microwave-excited and are described in detail in section 5.2.1. Such lamps are far more intense than hollow cathode lamps, but more difficult to operate with equivalent stability. **Radiofrequency excited** electrodeless discharge lamps (the radiofrequency region extends from 100 kHz to 100 MHz, whereas the microwave region lies around 100 MHz) are typically less intense (only 5–100 times **more intense** than hollow cathode lamps), but more reproducible. Commercially available radiofrequency lamps have a built-in starter,[1] are run at **27 MHz** from a simple power supply (capable of supplying 0–30 W), pretuned and **enclosed** to stabilize the temperature and hence the signal.

A diagram of such a lamp is shown in Fig. 4.3, taken from Barnett *et al., At. Absorpt. Newsl.* **15**, 33 (1976). This paper gives a good account of the analytical performance of electrodeless discharge lamps.

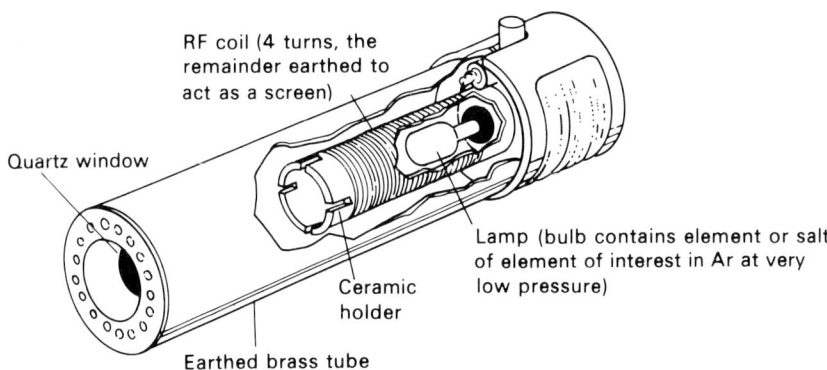

Fig. 4.3 Cutaway diagram of an r.f. excited electrodeless discharge lamp.

High intensity is not a source requirement in AAS and therefore electrodeless discharge lamps are unlikely to replace hollow cathode lamps, except for those elements which produce poor hollow cathode lamps. In this case, the **signal-to-noise ratio**, because of the low signal, may so adversely affect detection limits that electrodeless discharge lamps can offer improvements. The analytical signal is a ratio of I to I_o. Therefore, improved intensity of the signal can never improve sensitivity in AAS. (This rather subtle, but important, point will perhaps become clearer after sensitivity and detection limits have been discussed in sections 4.2.5 and 7.2.)

---

[1] The starter provides a high-voltage spark to ionize some of the filler gas for initiation of the discharge.

AN INTRODUCTION TO ATOMIC ABSORPTION SPECTROSCOPY

Source requirements in AAS

This leads us to a summary of source requirements in AAS. The source must give a **narrow resonance line profile** with little background, and should have a **stable** and **reproducible** output of sufficient intensity to ensure high **signal-to-noise** ratios. The source should be **easy to start**, have a **short warm-up** time and a long shelf life.

Q. How are the metal atoms produced and excited in a hollow cathode lamp?

Q. What is the normally preferred filler gas in a hollow cathode lamp?

Q. Why must quartz windows be used in sources for AAS?

Q. What are the advantages of radiofrequency excited electrodeless discharge lamps?

Q. Why does greater source intensity not lead to increased absorbance?

4.2.2 The most popular atom cells: flames

Several types of atom cell have been used for AAS. Of these, the most popular is still the flame, although a significant amount of analytical work is now performed using various electrically-heated graphite atomizers. This second type of atom cell is dealt with at length in Chapter 11, and the material here is confined to flames which have already been introduced in Chapter 2. (The reader is particularly reminded of the discussion in section 2.4 of the processes in flames.)

In AAS, the flame is only required to produce **ground state atoms** (cf. AES, where a hot flame is preferred as atoms must also be excited). Frequently, an **air–acetylene** flame is sufficient to do this. For those elements which form more refractory compounds, or where interferences are encountered (see section 6.2), a **nitrous oxide–acetylene** flame is preferred. In either case, a slot burner is used (100 mm for air–acetylene, 50 mm for nitrous oxide–acetylene) to increase the **path length** (this arises from Eqn (4.3), section 4.1) and to enable a specific portion of the flame to be viewed. Atoms are not uniformly distributed throughout the flame and, by adjusting the burner up and down with respect to the light beam, a **region of optimum absorbance** can be found.

FLAME ATOMIC ABSORPTION SPECTROSCOPY

Q. What are the requirements of a flame in (i) AES; (ii) AAS?

Q. Why are long slot burners preferred for AAS?

4.2.3 Instrument design

Most modern instruments use **gratings** rather than prisms for dispersion; Czerny–Turner, Ebert and Littrow systems are employed. As only **moderate resolution** is needed (because of the 'lock and key' effect), focal lengths of 0.25–0.5 m and rulings of 600–3000 lines mm^{-1} are commonly employed. Resolutions in the region 0.2–0.02 nm are claimed. The Annual Reports on Analytical Atomic Spectroscopy (see Appendix C) contains a full table of commercially available atomic absorption instruments each year.

As atomic emission and atomic absorption take place at the **same wavelength**, it is useful to be able to discriminate between the two signals to maximize atomic absorption sensitivity. Figure 4.4 shows how this may be

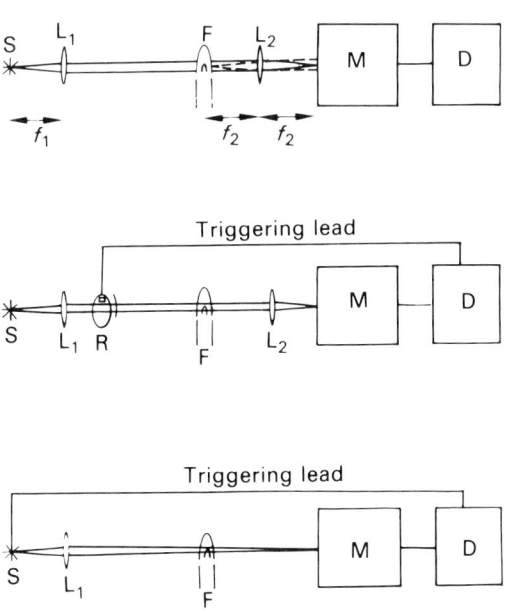

Fig. 4.4 Optical and instrumental arrangements for AAS. Key: S = source; L_1 = first lens; L_2 = second lens; F = flame; f = focal length; R = rotating sector; M = monochromator; D = detector.

done. In Walsh's first design (Fig. 4.4.1), the light from the lamp passes through the flame (the source being at the focus of the first lens) in a parallel beam which is focused on the entrance slit of the monochromator, this being at the focus of the second lens. The flame is placed at the focus of the second lens, so that flame emission is exactly defocused at the monochromator. Thus, the atomic absorption signal is maximized and the atomic emission signal minimized.

The next development was **modulation**. A **rotating sector** (often crudely referred to as a chopper) is placed in the light beam (Fig. 4.4.2). As the beam strikes the solid part of the chopper, it is interrupted; the hole in the sector allows it to pass. The sector is placed between the source and the flame. The atomic absorption signal is now modulated, but the atomic emission signal is not. An **AC amplifier** tuned to the atomic absorption signal, via **phasing coils** on the rotating sector, **gates** the amplifier and selectively amplifies the atomic absorption signal as opposed to the DC atomic emission signal. It is thus essential that the sector be placed between the source and the flame. In atomic absorption and atomic emission spectrometers, two rotating sectors may be found, one for atomic absorption and one for atomic emission (between the flame and the monochromator). This atomic emission chopper only functions when the source is off and is necessary if the instrument is fitted with an AC amplifier. When the instrument is switched back to atomic absorption mode, the sector should lodge in the open mode.

Often flame radiation may be reflected from the back of a rotating sector, which is less than totally reliable because it is mechanical. Many modern instruments therefore **modulate the power** applied to the source and also use this signal to **trigger the amplifier** (Fig. 4.4.3). Such instruments may still have a rotating sector (but only for atomic emission) between the flame and the monochromator, or the amplifier may be capable of being reset for signals.

In all cases, a photomultiplier is used as the detector, and, after suitable amplification, a variety of read-out devices may be employed (see section 4.2.4).

Double beam instruments

The systems so far described have all been **single-beam** spectrometers. As in molecular spectrometry, a double-beam spectrometer can be designed. This is shown diagrammatically in Fig. 4.5. The light from the source is split into **two beams**, usually by means of a rotating **half-silvered mirror** or by a **beam splitter** (a 50%-transmitting mirror). The second reference beam passes behind the flame and, at a point after the flame, the two beams are recombined. Their ratio is then **electronically compared**.

Double-beam operation offers far fewer advantages to AAS than it does to molecular absorption spectrometry, mainly because the reference beam

FLAME ATOMIC ABSORPTION SPECTROSCOPY

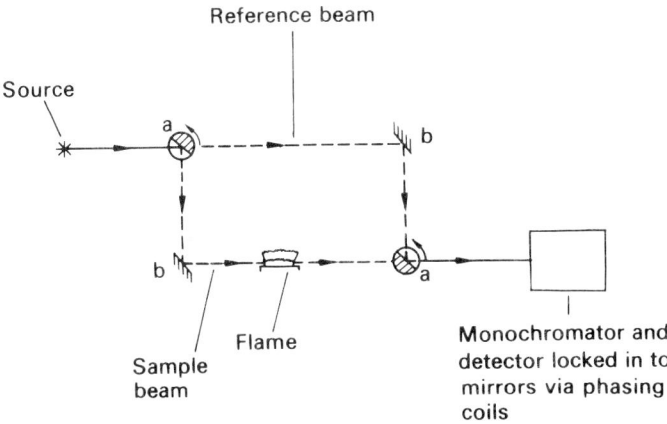

Fig. 4.5 Double beam atomic absorption instrumentation. (a) Rotating half-silvered mirror; (b) front surface mirror.

does not pass through the most noise-prone area of the instrument, the flame. Double-beam systems can compensate for **source drift**, warm-up and **source noise**. This should lead to improved **precision** and often does. However, as the major source of noise is likely to be the flame, this advantage is slight and may be more than offset by the significant loss of intensity in the light signal, and hence lower signal-to-noise ratio.

Background correction

Considerably more advantage can be derived from the use of a second beam of **continuum radiation** to correct for non-atomic absorption. Figure 4.6 schematically summarizes how this operates. When using a **line source** such as a hollow cathode lamp, we observe **atomic absorption** in the flame, **absorption from molecular species** and scattering from particulates. The latter, known as **non-specific absorption**, is a particular problem at shorter wavelengths and can lead to positive errors. When using a **continuum source** (e.g. a deuterium arc or a hydrogen hollow cathode lamp), the amount of atomic absorption observed, as we have already seen (section 4.1), is negligible, but the **same amount of non-specific** absorption is seen. Thus, if the signal observed with the continuum source is **subtracted** from that observed with the line source, the error is removed.

Figure 4.7 shows an instrument capable of doing this **simultaneously and automatically**. Lead is particularly prone to this problem, and Fig. 4.8 shows how background correction can be used to remove the interference of non-specific absorption when determining lead in chromium. Notice that the **precision** is improved too, mainly because the effects which give rise to the background are not very reproducible.

AN INTRODUCTION TO ATOMIC ABSORPTION SPECTROSCOPY

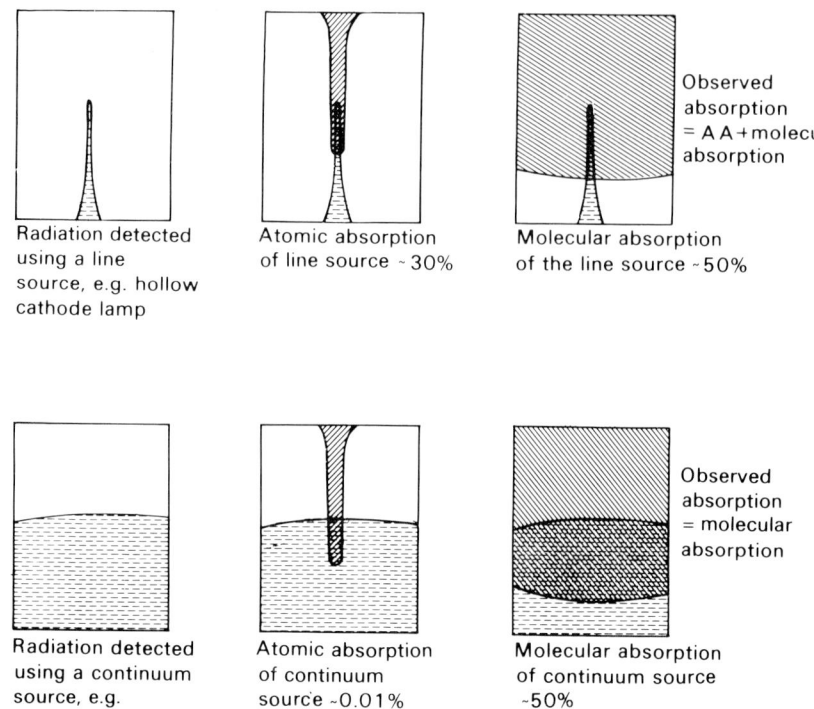

Corrected atomic absorption =
Absorption observed using line source − absorption observed using continuum source

Fig. 4.6 Background correction.

A recent development has been the use of the **Zeeman** effect for background correction. An atomic spectral line, when generated in the presence of a strong **magnetic field**, can be split into a number of components of slightly different wavelength. The field can be applied to either the source or the atom cell. In the normal Zeeman effect, the line appears as **three components** when viewed perpendicularly to the magnetic field (see Fig. 4.9). The π **component** is situated at the 'normal' wavelength of the line, but the σ^+ and σ^- components lie an equal distance on either side. The σ components are **linearly polarized** perpendicular to the magnetic field. If the field is strong enough, these components will lie outside the atomic absorption profile and background can be corrected by measuring the absorbance of the π and σ components separately.

Newstead, Price and Whiteside have published an excellent **review** on background correction (*Prog. Anal. At. Spectrosc.* **1**, 267 (1978)). It should be noted, however, that there have been advances in the format of Zeeman systems since this review was published.

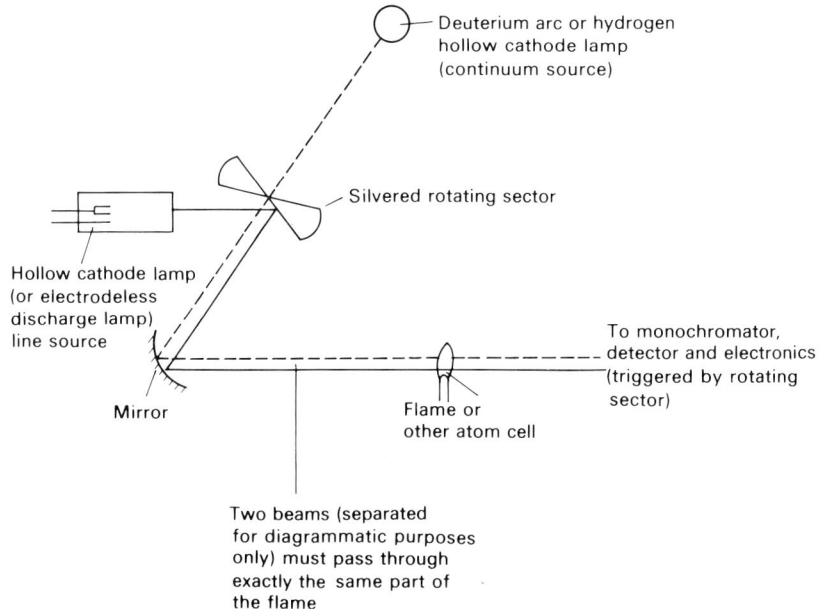

Fig. 4.7 Automatic simultaneous background corrector.

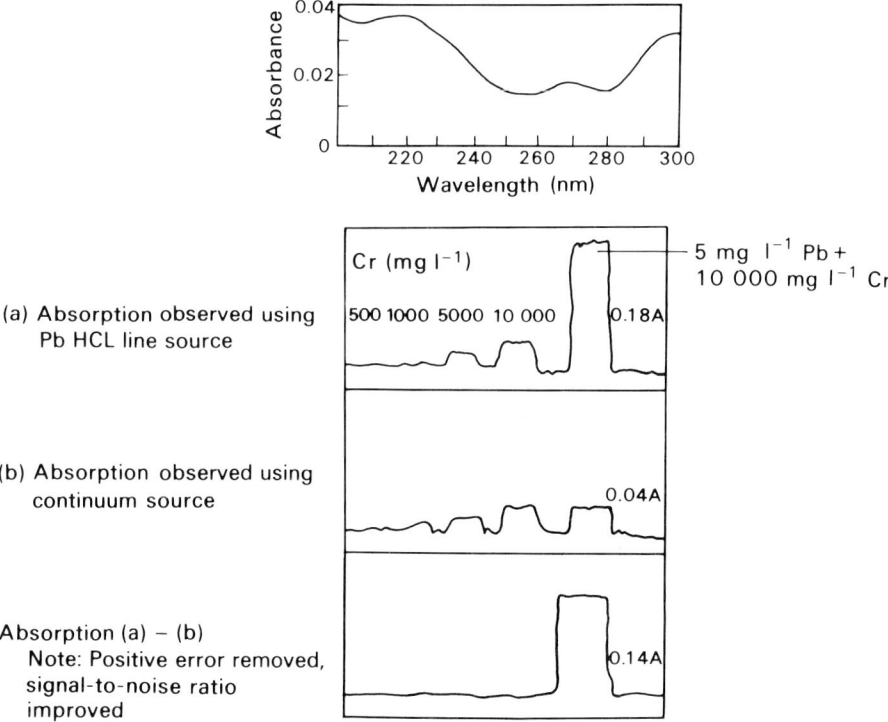

Fig. 4.8 Background correction in action. (Top) background due to spraying chromium blank. (Bottom) determination of Pb in Cr.

AN INTRODUCTION TO ATOMIC ABSORPTION SPECTROSCOPY

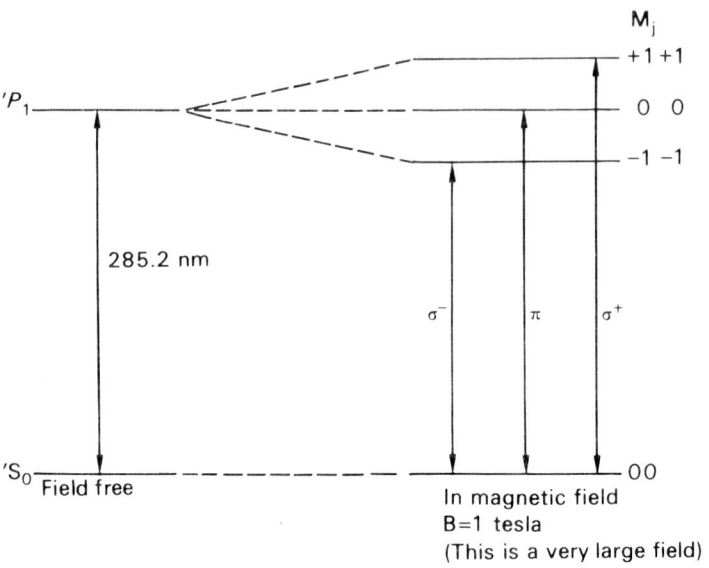

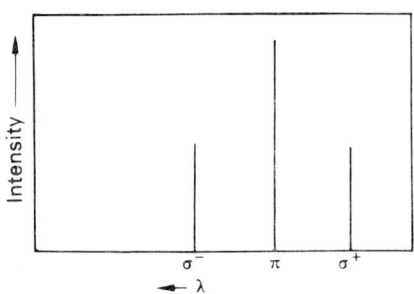

(The diagram exaggerates the separation of components for clarity. In this instance, it would be 0.0038 nm.)

Fig. 4.9 Normal Zeeman effect for magnesium. A readable review of the Zeeman effect by Brown can be found in *Anal. Chem.*, **49**, 1269A (1977).

Q. How can the interference of AA measurements by AE signals be removed?

Q. What are the advantages in AAS of (i) modulation; (ii) double-beam systems; (iii) background correction?

FLAME ATOMIC ABSORPTION SPECTROSCOPY

4.2.4 Read-out systems

After suitable **amplification**, the signal may be read out on an **analogue** meter. The lamp, electronics and **flicker** of the flame may all contribute to short-term, **irregular fluctuations** in the signal, i.e. noise. The meter will follow these to some extent, and flicker on the needle will not help the observer. It is usual to include some electronic **damping** to reduce the noise, i.e. to average it out to some extent. The damping must not be too large or the meter will be sluggish, i.e. the **time constant** should not be greater than 2 s. If an accurate picture of the noise is desired, a **charter-recorder** with fast-response (e.g. 1 s for full scale deflection) and low damping should be used.

Modern instruments usually present the results on a **digital meter**. It is well to remember that such a meter can only display a single reading and, rather than display instantaneous readings at the moment the display is updated, it is preferable to use **integration**. The signal is continuously stored, or integrated, in a condenser for a fixed (usually selectable) period of time. After normalization, the integrated signal is displayed. Such integrating circuits are a useful addition to an analogue meter, as they offer a more rapid way of reducing noise.

The photomultiplier output is **proportional to the transmission** of the flame, yet concentration is **proportional to absorbance**. A **logarithmic** read-out is preferred as being **linear** in absorbance. This can be achieved by using an appropriate logarithmic amplifier, or when normalizing the output from the integrating condenser.

Most instruments offer **scale expansion** facilities which are essential for **accurate work** at low or high concentrations. The signal can be electrically expanded (as is the **noise**) to a fixed factor, or continuously, so that the reading of a chosen standard comes to a desired figure (e.g. 10 ppm reads 100). This latter option is referred to as **'concentration read-out'**. Care must be taken when using this at high concentrations, as few calibration curves are still linear above 0.5 absorbance units. Curves tend to **bend towards** the concentration axis, principally because **stray light** (i.e. unabsorbable light, for example from nearby lines in the lamp) becomes an important contribution to the total light falling on the detector. Manual controls for **'curve correction'** may be supplied which enable the calibration to be **linearized** by extra expansion on standards of high concentration.

In modern instruments, curve correction and other read-out facilities may be handled by a built-in **microprocessor**. The correction system has to compute the most reliable curve on which the blank and standards fit. In some instruments, a **microcomputer** may actually control instrumental settings, and in response to a command to determine a particular element, will move at high speed to the **correct wavelength**, search out the peak, call for calibration standards (usually supplied by an attentive human assistant!) and perhaps also control other features, such as the flame composition and height. In less expensive instruments, the microprocessor merely controls

AN INTRODUCTION TO ATOMIC ABSORPTION SPECTROSCOPY

the **computation** and **read-out**. In this case, curve correction can be carried out in a more theoretically valid way using a number of standards, an equation for stray-light effects and standard computerized curve-fitting techniques. If the instrument is told, via **'typed in'** instructions, the concentration of the standards and which are the unknowns, a print-out can be obtained concerning the concentration of each unknown and the **confidence limits** (or standard deviation). The microprocessor is supplied with a **keyboard** and permanently-programmed ROMs (**read only memories**), in which the manufacturer has stored details of how to do the calculations. Using the keyboard, information on the concentration of standards and the solution being sprayed can be given to the RAMs (**random access memories**), which can be entered and cleared when the user wishes. The result may be printed on to **tape** or presented on a **visual display unit**, or both.

Q. In AAS, where does noise come from?

Q. How can noise problems be minimized?

Q. What is the advantage of logarithmic read-out in AAS?

Q. When is curve correction needed and how can it be performed?

Q. What is the difference between ROMs and RAMs?

4.2.5 Sensitivity and limit of detection

In section 4.1, we noted that the source line profile should be **narrower** than the absorption line profile. In section 2.7, we saw that **broadening** is related to temperature and pressure. Now, the temperature in a hollow cathode lamp or an electrodeless discharge lamp is a few hundred degrees centigrade and the pressure is about 5 torr. In a flame, the temperature is 2000–3000 °C and the pressure is atmospheric (760 torr). Thus, the Walsh requirement is easily met.

The power of detection of AAS is conveniently expressed as the lower **limit of detection** of the element of interest. This l.o.d. is derived from the smallest measure x which can be accepted with **confidence** as genuine and is not suspected to be only an accidentally high value of the blank measure. The value of x at the 95% confidence level (so called **2σ level**) is given by

$$x = \bar{x}_{bl} + 2s_{bl}$$

FLAME ATOMIC ABSORPTION SPECTROSCOPY

where $\bar{x}_{bl}$ is the mean and s_{bl} is the estimate of **standard deviation**[1] of the blank measures. Thus, the limit of detection may be defined as 'that quantity of the element which gives rise to a reading equal to twice the standard deviation of a series of at least ten determinations at or near the blank level'. This assumes a 'normal' distribution of errors, and may consequently result in more or less optimistic values.

The limit of detection is a useful figure which takes into account the stability of the total instrumental system. It may vary from instrument to instrument and even from day to day as, for example, mains-borne noise varies. Thus, spectroscopists often also talk about the **characteristic concentration** (often erroneously referred to as the **sensitivity**—erroneously, as it is the reciprocal of the sensitivity) for **1% absorption**, i.e. that concentration of the element which gives rise to **0.0044 absorbance units**. This can easily be read off the calibration curve. The characteristic concentration is dependent on such factors as the atomization efficiency and flame system, and is independent of noise. Both this figure and the limit of detection give different, but useful, information about **instrumental performance**.

Q. Why does the 'lock and key' effect occur in flame AAS?

Q. Define limit of detection and characteristic concentration.

Q. What information is given by the limit of detection, and how does this differ from that given by the characteristic concentration?

---

[1] The **deviations** of a number of measurements from the **mean** of those measurements will show a **distribution** about the mean. If that distribution is **symmetrical** (or to be more precise **Gaussian**), this is termed a **normal error curve**. Thus, there is always some uncertainty in any measurement. The standard deviation is a useful parameter derived from the normal error curve. An estimate s of the true standard deviation σ of a finite set of n diffent readings can be calculated from:

$$s = \sqrt{\left(\frac{\text{sum of the squares of the difference between each reading and the mean}}{n-1}\right)}$$

Statistical theory tells us that, provided sufficient readings are taken, 68.3% of the actual readings lie within the standard deviation of the mean, and that the mean $\pm 2\sigma$ and the mean $\pm 3\sigma$ will contain 95.5% and 99.7% of the readings, respectively. Thus, there is only a 5% chance that a reading larger than the mean of the blank readings by twice the standard deviation is merely due to an unusually high blank reading.

5 FLAME ATOMIC FLUORESCENCE SPECTROSCOPY

Of the three techniques covered in this book, this is the least widely used, and our discussion of flame AFS will therefore be somewhat brief.

5.1 THEORY

The six **types** of flame AFS which have been observed are summarized in Fig. 5.1. The most intense AFS generally concerns **resonance fluorescence** and this has been the most useful analytical type. **Direct-line fluorescence** mechanisms have also been exploited analytically, as here the emission is at a wavelength different from that of the exciting radiation and, by using filters, the effects of **scattering** can be eliminated.

Kirkbright and Sargent, and Winefordner (see Appendix C) deal thoroughly with theories describing the fluorescence intensity when using line sources, continuum sources and lasers for excitation. We will only deal with the theory as it concerns **line sources**. The intensity of fluorescence, I_F, is **directly proportional to the source intensity**, I_s.

At low concentrations, I_F is proportional to the **concentration** of the analyte; at higher concentrations, self-absorption is observed (cf. AES). This behaviour is reflected in the **growth curves** of log I_F versus log N (one of the earliest practical confirmations of the theory given in the texts above can be found in Ebdon *et al.*, *Talanta* **17**, 965 (1970). Figure 5.2 is based on these results. These curves often show linearity of 4 orders of magnitude before curvature.

As I_F depends on I_s, a stable, **intense** sharp-line source greatly enhances AFS sensitivity. Similarly, the **geometry** of the atom cell is important.

Q. Which type of AFS is of greatest practical use?

Q. Can sensitivity be improved by increasing the intensity of the source in (i) AAS; (ii) AFS?

FLAME ATOMIC FLUORESCENCE SPECTROSCOPY

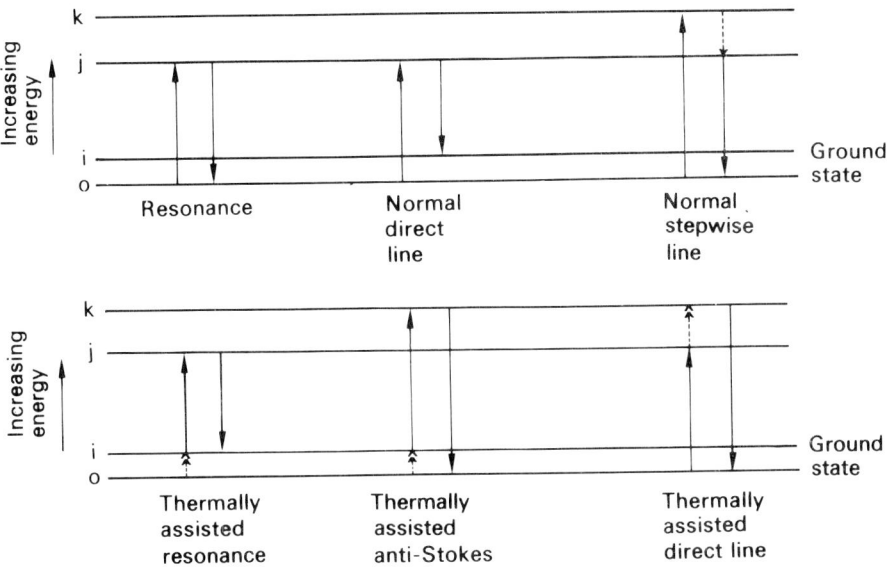

Fig. 5.1 Types of atomic fluorescence. The solid lines represent radiational processes, and the dashed lines non-radiational processes. In the latter, a single-headed arrow represents non-radiational deactivation and a double-headed arrow a thermal activation process. The term anti-Stokes is used when the radiation emitted is of shorter wavelength, i.e. greater energy than that absorbed.

5.2 INSTRUMENTATION

Only one commercial instrument for flame AFS has ever been marketed, and that only for a short time. However, flame AFS may be performed simply on a modified atomic absorption spectrometer. Some of the components, which have been described in the many published papers, are described below.

5.2.1 Sources

A source for AFS must be very intense to give low detection limits. In contrast with AAS, **continuum sources** may be used, but, given the distribution of wavelengths from a black-body radiator, few give high intensity in the vital ultraviolet region. **Xenon arcs** have been used for AFS, but scatter problems are encountered. The actual integrated intensity over the absorption half-width is still relatively low, whereas for a line source the majority of the emission is concentrated in a few lines. Hence, the preference is for **line sources**.

Where **vapour discharge lamp** sources exist (for Hg, Na, Cd, Ga, In, Tl and

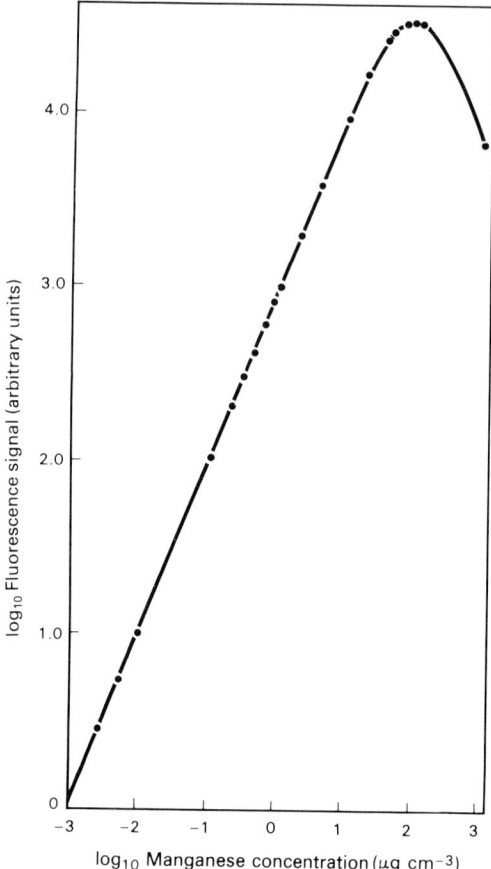

Fig. 5.2   Growth curve for manganese atomic fluorescence at 280 nm.

Zn) they can be used. **Hollow cathode lamps** are insufficiently intense, unless operated in a **pulsed** mode. **Microwave-excited electrodeless discharge lamps** are very intense (typically 200–2000 times more intense than hollow cathode lamps) and have been widely used. They are inexpensive and simple to make and operate, but stability has always been the main problem with this source. Recently this has been greatly improved by operating the lamps in microwave cavities **thermostatted** by warm air currents. A typical electrodeless discharge lamp is shown in Fig. 5.3. This type has been made on numerous occasions by the author using a simple vacuum line, quartz tubing and an oxygen–natural gas flame.

The high intensity of **lasers** makes them obvious, but expensive, candidates as sources for AFS. The advent of the **tunable dye-laser** led to the possibility of selecting wavelengths from a laser, and **frequency doubling** (second harmonic generation) has enabled the excitation of AFS lines in the ultraviolet region. The light from such a source can be sufficiently intense

FLAME ATOMIC FLUORESCENCE SPECTROSCOPY

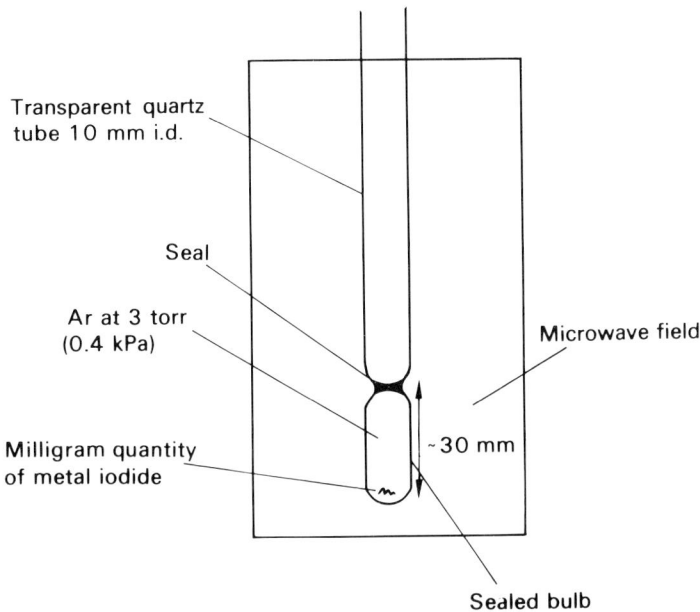

Fig. 5.3 A microwave-excited electrodeless discharge lamp.

to cause **saturation fluorescence** (i.e. population inversion), thus nullifying the effects of quenching (e.g. radiationless return to the ground state induced by flame species at flame temperatures) and self-absorption.

Q. Why are continuum sources rarely used in (i) AAS; (ii) AFS?

Q. What advantages are offered to AFS by (i) electrodeless discharge lamps; (ii) lasers?

5.2.2 Flames

The **fluorescence power yield** is always less than one. This non-radiative loss of energy is referred to as **'quenching'**. Quenching increases with temperature (the number of collisions) and quenching **cross-section** of the colliding particle (argon has a negligible quenching cross-section, hydrogen a low one, oxygen a high one). The ideal atom cell for AFS would also exhibit no background, thus enabling the detection of very small signals.

There has been interest in the low radiative background, low quenching

AN INTRODUCTION TO ATOMIC ABSORPTION SPECTROSCOPY

argon–hydrogen diffusion flame. The temperature of this flame is too low to prevent severe chemical interferences and therefore the **argon separated air–acetylene** flame has been most widely used. The hot nitrous oxide–acetylene flame (argon separated) has been used where atomization requirements make it essential. In all cases, **circular flames**, sometimes with mirrors around them, offer the preferred geometry.

Q. Why does the sensitivity of AFS decrease in the order argon–hydrogen; air–acetylene; nitrous oxide–acetylene, whereas that for AES increases in the same order.

5.2.3 Instrument design

In some flame AFS systems, **interference filters** and **solar blind photomultipliers** have been used, but usually a conventional **monochromator** is used. As in AAS, the source signal is **modulated** so that the AFS can be distinguished from the AES.

It has been shown that a **high frequency** of modulation of the electrodeless discharge lamp (e.g. 10 kHz) is advantageous. This frequency is well away from the low frequency of flame noise. If the amplifier is **'locked-in'** to this high frequency via a **reference signal**, optimum signal-to-noise ratio is achieved.

Q. Why is modulation used in AFS?

---

6 INTERFERENCES AND ERRORS IN FLAME SPECTROSCOPY

Interference is defined as an effect causing a **systematic deviation** in the measure of the signal when a sample is nebulized, as compared with the measure that would be obtained for a solution of equal analyte concentration in the same solvent, but in the absence of **concomitants**. The interference may be due to a particular concomitant or to the combined effect of several concomitants. A concomitant causing an interference is called an **interferent**. Interference only causes an **error** if not adequately corrected for during an analysis. Uncorrected interferences may lead to either **enhancements** or **depressions**. Additionally, errors may arise in analytical methods in other ways, e.g. in **sample pre-treatment**, via **operators** and through the **instrumentation**. We shall deal first with errors and then look at interferences in some detail.

6.1 SAMPLE PRE-TREATMENT ERRORS

Obviously, the **accuracy** of an analysis critically depends on how **representative** the sample is of the material from which it is taken. The more **heterogeneous** the material, the greater care must be taken with sampling. The analytical methods described in this book can typically be used on small samples (e.g. 100 mg of solid or 10 cm^3 of liquid), and this again heightens the problem. Readers are referred to a general analytical text for details on sampling, but it should be stressed that if either the concentration of the analyte in the sample does not represent that in the **bulk material,** or the concentration of analyte in the solution at the time it is presented to the instrument has changed, the resultant error is likely to be greater than any other error discussed here. Regrettably, the supreme importance of these points is not always recognized.

Usually, liquid samples are sprayed into the flame. Thus, solid samples must be **dissolved**. **Analytical grade reagents** must be used for dissolution to prevent **contamination** at trace levels. Certain volatile metals (e.g. zinc and cadmium) may be lost when **dry ashing** and volatile chlorides (e.g. arsenic and chromium) lost upon **wet digestion**. It is particularly easy to lose mercury during sample preparation. Appropriate steps must be taken in the choice of method of dissolution, acids and conditions (e.g. whether to use reflux conditions) to prevent such losses. Trace metals may also be lost by **adsorption** on to precipitates, such as the silica formed on digestion using oxidizing acids. This possibility should be investigated (e.g. by recovery tests).

Glass-ware may give rise to further errors. Adsorption is a particular problem at trace levels. Few solutions below a concentration of 10 ppm can be considered **stable** for any length of time. Various **preservatives** to guard against adsorption of metals on to glass-ware have been reported in the literature. Common precautionary steps are to keep the **acid concentration** high and to use **plastic laboratory ware**. Laboratory ware may contaminate samples (e.g. sodium and silicon from glass, copper and zinc from polythene) and contamination may also occur from **airborne particulates** (e.g. lead). As a precaution, blanks should always be run.

Q. Under what conditions is sampling most problematical?

Q. How can we minimize the possibilities of (i) contamination; (ii) losses by adsorption?

6.2 OPERATOR ERRORS

Experience tells us that no account of possible operator errors can ever be exhaustive. The use of a **line source** and the **ratio method** (i.e. ratio of I_o/I) tend to minimize errors in AAS. Thus, if the **wavelength setting** is seriously incorrect for AAS, it is unlikely that any absorption will be observed at all. If the wavelength is incorrectly tuned, the effects on the value of I will roughly equal those on the value of I_o, and the error may not be too serious. AES does not share this advantage, as the flame will emit at many wavelengths and AES is not usually a ratio technique.

A frequent source of error is **poor standards**. Besides the obvious error of standards wrongly made, it should not be forgotten that trace metal standards are **unstable**. Concentrations of 10 ppm and less usually need to be prepared daily. Even standards purchased from commercial suppliers will

INTERFERENCES AND ERRORS IN FLAME SPECTROSCOPY

age and this is especially true when chemical changes can be expected in the analyte (e.g. silicon).

Readings should only be taken when the signal has reached **equilibrium**. When a new sample is presented, several seconds will elapse before the **cloud chamber** reaches equilibrium. This equilibration time will increase if **damping** is being used in the instrument. Particular care must be taken when the concentration of samples or standards changes markedly, especially if the new solution is more **dilute**.

Q. Why are errors associated with monochromator settings less in AAS than in AES?

Q. How can poor standards introduce analytical errors?

Q. Why should the integrate button not be pressed until several seconds after a sample change?

6.3 INSTRUMENT ERRORS

These are usually more acute in AFS and AES than in AAS. This is because AAS is a **ratio method** and many instrumental errors (e.g. long-term source drift, small monochromator drifts) should cancel out, as I is ratioed to I_o.

For AES, a **stable flame temperature** is vital (see Eqn (3.5), section 3.1). Thus, draughts should be avoided and solvent always aspirated (water exerts a considerable cooling effect on the flame). Flame gas controls should not drift.

For AFS, **source noise and drifts** give the greatest problems, as the source is not monitored directly.

In both AES and AFS, **monochromator drift** and slit-width drift often caused by the flame temperature, can bring about errors.

In all three techniques, a stable uptake rate, or **aspiration rate**, is required. This falls as the viscosity of the solution sprayed is increased. Nebulizer-uptake interferences can be minimized if the **dissolved salts** content of samples and standards is approximately matched. For example, when determining ppm sodium levels in 2 M phosphoric acid, ensure that the standards are also dissolved in 2 M phosphoric acid, using a blank to check for contamination.

AN INTRODUCTION TO ATOMIC ABSORPTION SPECTROSCOPY

> **Q.** Why are different instrumental errors encountered in AES, AAS and AFS?

> **Q.** Why is the observed absorbance of 2 ppm zinc dissolved in dilute hydrochloric acid greater than that of 2 ppm zinc dissolved in molar potassium dichromate solution?

6.4 INTERFERENCES

We will now deal with errors which may be caused by interferents. All the interferences described require the presence of analyte, except the first type—spectral interferences.

6.4.1 Spectral interferences

The nature and extent of spectral interferences in AES are greatly **different** from those in AAS and line-source AFS.

In AES, the requirement is that the **monochromator should separate** the line of interest from other **lines and bands** emitted from the flame (e.g. the Mn 403.076 nm emission line from the Ga 403.298 nm line and the K 404.414 nm line, or the Mg 285.21 nm line from the Na 285.28 nm line). Given the typical resolution of 0.2 nm noted with AAS monochromators used in AES (section 4.2.3), spectral interferences occur in AES (see section 3.2 for ways of reducing such problems). Fortunately, the lithium, sodium and potassium lines are well separated, but there are problems with a calcium hydroxide band at the sodium line, and a strontium hydroxide band at the lithium line.

In AAS and line-source AFS, the problem of spectral interference is much reduced and **line overlap interferences are negligible**. This is because the resolution is provided by the 'lock and key' effect. To give a spectral interference, the lines must not merely be within the band pass of the monochromator, but actually overlap each other's **spectral profile** (i.e. be within 0.01 nm). West (*Analyst* **99**, 886 (1974)) has reviewed all the reported (and a number of other) 'spectral interferences' in AAS. Most of them concern lines which would never be used for a real analysis, and his conclusion is that the only 'real' problem is in the analysis of copper heavily contaminated with europium! The most commonly used copper resonance line is 324.754 nm (characteristic concentration, 0.1 ppm) and this is overlapped by the europium 324.753 nm line (characteristic concentration, 75 ppm).

Spectral interferences from the overlap of **molecular bands and lines** (e.g. the calcium hydroxide absorption band on barium at 553.55 nm) cannot be

INTERFERENCES AND ERRORS IN FLAME SPECTROSCOPY

so easily dismissed for AAS. Lead seems to be particularly prone to such **non-specific absorption** problems at the 217.0 nm line (e.g. sodium chloride appears to give strong molecular absorption at this wavelength). This type of problem is encountered in practical situations, but can be removed by the technique of **background correction** (see section 4.2.3.).

Q. Why can it be said that spectral interferences can be virtually eliminated in AAS, but pose the severest limitation in AES?

Q. How can non-specific absorption problems in AAS be overcome?

6.4.2 Ionization interferences

This is a **vapour phase interference**, previously referred to as **cation enhancement**. In the air-acetylene flame, the intensity of rubidium emission or absorption can be doubled by the addition of potassium. This is straightforward **ionization suppression** (see section 2.4), but if uncorrected will lead to substantial positive errors when the samples contain easily ionized elements and the standards do not; for example, when river water containing varying levels of sodium are to be analysed for a lithium tracer, and the standards, containing pure lithium chloride solutions, do not contain any ionization suppressor.

The problem is easily overcome by adding an **ionization suppressor** (or **buffer**) in large amount to all samples and standards.

Q. Why is ionization interference equally severe in AAS, AES and AFS?

Q. What precautions with regard to both standards and samples would be needed when determining potassium in sea-water?

6.4.3 Chemical interferences

Given how easily the two types of interference discussed above can be overcome, this third type constitutes the biggest source of problems in AAS. A brief discussion is given of **solid phase interferences** centred around the following **classification**:

(a) depressions caused by the formation of less-volatile compounds which are difficult to dissociate;
(b) enhancements caused by the formation of more-volatile compounds;
(c) depressions due to occlusion into refractory compounds;
(d) enhancements due to occlusion into more-volatile compounds.

AN INTRODUCTION TO ATOMIC ABSORPTION SPECTROSCOPY

(a) Formation of less-volatile compounds

The most famous of this type of interference is that of **phosphate on calcium**; sulphate and silicate have a similar effect. Figure 6.1.1 illustrates the

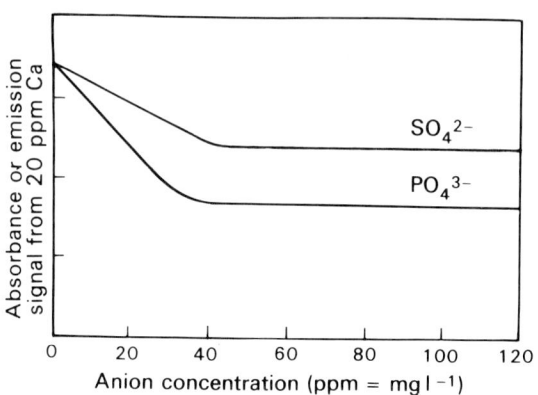

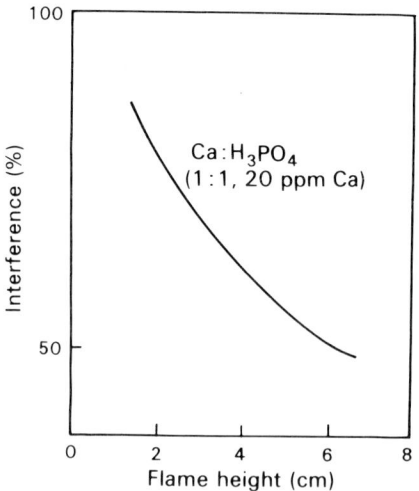

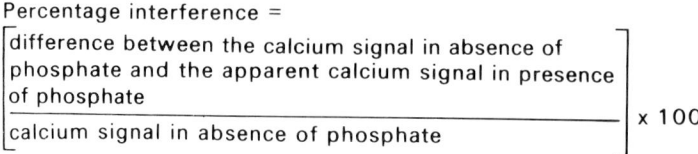

Fig. 6.1 Anion interferences on calcium.

effect of increasing phosphate concentration on the calcium signal. The graph shows a pronounced **'knee'** at ratios of P:Ca variously reported in a range 0.3–1.1. The effect is less pronounced higher up the flame (see Fig. 6.1.2), and is not observed in a hotter flame, such as the nitrous oxide–acetylene flame. The effects are the same in AAS, AES or AFS, at the 422.7 nm or 620.0 nm line, or the calcium hydroxide band at 554 nm. The **constant level** of interference observed above a certain anion level strongly suggests formation of a compound (probably a calcium phosphate). This compound is less volatile than calcium chloride, and hence the formation of calcium atoms is hindered.

There are many examples of this type of interference, several involving the formation of **aluminates**. They should all show the pronounced 'knee' in the graph—this distinguishes them from non-specific occlusions.

Several **approaches** can be made **to reduce** such interferences.

(i) Use a **hotter flame**.
(ii) **Adjust the nebulizer** to produce a smaller particle size.
(iii) Make **observations higher** in the flame.
(iv) Use a **'releasing agent'**, an element which will enter into a 'law of mass action competition' with the analyte to combine with the interferent. If an excess of the releasing agent is added, the analyte is released from the interfering anion (e.g. excess lanthanum or strontium releases calcium from phosphate interference).
(v) **Use a protective chelating agent**, which preferentially complexes the analyte, protecting it from the 'grasp' of the interferent (e.g. excess EDTA protects calcium from phosphate interference).

(b) Formation of more-volatile compounds

These interference effects are far less common. Under this heading, some authors classify the **enhancement** of signals from several, otherwise refractory elements by fluoride. The use of protective agents (e.g. EDTA for calcium, or 8-hydroxyquinoline for aluminium or chromium) are also examples of this type of effect.

(c) Occlusion into refractory compounds

Such depressions can be encountered when the matrix is refractory (e.g. zirconium, uranium or a rare-earth element), and the small amount of analyte can be physically trapped in clotlets of matrix oxide in the flame. Such systems do not show a 'knee' (see (a)) and can be minimized by higher flame temperature.

(d) Occlusion into volatile compounds

Some compounds (e.g. ammonium chloride) explosively **sublime** in the

AN INTRODUCTION TO ATOMIC ABSORPTION SPECTROSCOPY

flame, thus enhancing atomization. By adding excess ammonium chloride to all samples and standards, this effect can also be used to minimize interferences of type (a) and (c).

Q. How can specific and non-specific depressions (types (a) and (c)) be distinguished?

Q. List several ways in which the interference of phosphate on calcium may be minimized.

Q. How might aluminium interfere with magnesium during AAS?

Q. Iron depresses atomic absorption by chromium in the air–acetylene flame. For what reasons do workers add 8-hydroxyquinoline and/or ammonium chloride to minimize this interference?

---

7 THE PRACTICE OF FLAME SPECTROSCOPY

All three spectroscopic techniques require **calibration** with standards of known analyte content. Fortunately, AAS is sufficiently specific for a simple solution of a salt of the analyte in dilute acid to be used, although it is a wise precaution to **buffer** the standards with any salt which occurs in large concentration in the sample solution, e.g. 500 ppm (1 ppm $=1$ mg l$^{-1}=1$ µg ml^{-1}) or above. **Calibration curves** for AES, AAS and AFS can be obtained by plotting, respectively, **emission** signal, **absorbance** (not percentage transmission) or **fluorescence** signal against **concentration**. Often the calibration curve will **bend** towards the concentration axis at higher concentrations. In AES and AFS, the most likely cause will be **self-absorption** (see sections 3.1 and 5.1), while in the case of AAS, the cause will probably be **stray, or unabsorbable, light** (see section 4.2.4). As the slope decreases, so will **precision**, and it is preferable to work on the **linear portion** of the calibration, the **working curve**. The most common cause of **upward curvature** is self-suppression of **ionization**, and this can be cured by adding an ionization suppressor (see section 6.4.2). Best results are obtained when the standards are sprayed first in ascending order of concentration, and then in descending order, each time **'bracketing'** the samples with standards of immediately lower then higher concentration when ascending, and the reverse when descending. Modern instruments will often provide facilities for effectively performing calibration on the instrument, either by **'scale expansion'** or using a **microprocessor**. In the latter case, it is more important than ever that the sample should **not lie off** the calibration curve, i.e. it must never be more concentrated than the strongest standard, and preferably not more dilute than the weakest.

The method of **standard additions** is a useful procedure for checking the accuracy of a determination and overcoming interferences when the composition of the sample is unknown. It should be noted that the method cannot correct for **spectral interferences** and **background absorption**. At least three aliquots of the sample are taken. One is left untreated; to the others, **known additions** of the analyte are made. The additions should preferably

AN INTRODUCTION TO ATOMIC ABSORPTION SPECTROSCOPY

be about $\frac{1}{2}x$, x and $2x$, where x is the concentration in the unknown. The solutions are then aspirated and the curve shown in Fig. 7.1 plotted. The curve is **extrapolated** back until it crosses the x-axis, giving the concentration in the unknown. Correction for dilutions may also have to be made. On modern instruments, the procedure may be simplified as follows:

(i) spray the sample solution and auto-zero;
(ii) spray the strongest addition and set the concentration read-out display to that of the addition;
(iii) check linearity with the other additions;
(iv) spray the blank solution, and the negative value on the display will give the concentration in the original sample.

A microprocessor can be programmed to do this.

A standard addition curve **parallel** to the calibration curve is indicative (but not conclusive) of the absence of interference.

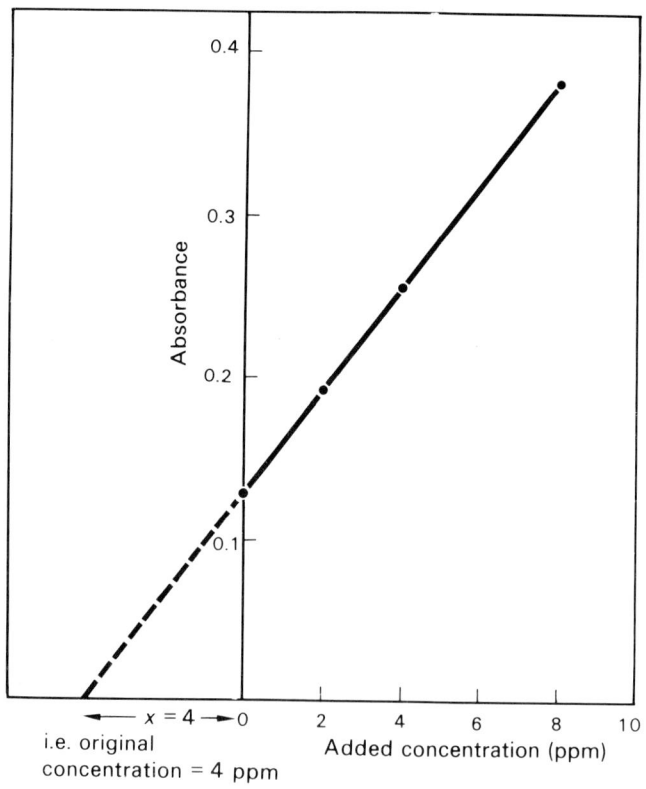

Fig. 7.1 Method of standard additions.

7.1 APPLICATIONS

Several of the texts listed in Appendix C are particularly informative on applications, and the applications tables published every year in the Annual Reports on Analytical Atomic Spectroscopy (see Appendix C) offer invaluable help in the development of new applications in the laboratory. Nearly all the applications of flame spectroscopy require the sample to be in **solution**. A list of **detection limits** for flame AAS is given (Table 7.1). Where possible, samples should be brought into solution to give analyte levels of at least ten times the limit of detection.

7.1.1 Clinical, food and organic samples

It is usually necessary to **destroy the organic material** before spraying the sample. Care must be taken to avoid **losses** of volatile elements (an oxidizing **wet ashing** procedure is preferred for elements such as lead, cadmium and zinc) and **contamination** from reagents. The Analytical Methods Committee of the Royal Society of Chemistry prefers a hydrogen peroxide–sulphuric acid (1:1) mixture for situations requiring strong attack on carbonaceous material, but clearly such a mixture must be treated with respect. Beverages must be degassed before spraying. In serum analyses, the protein is often precipitated with trichloroacetic acid before analysis, but only if the analyte is not likely to be co-precipitated. Direct aspiration of diluted and buffered samples is to be preferred.

Petrochemicals

As organic solvents are aspirated more efficiently (see section 2.4), aqueous standards cannot be used for calibration when determining trace metals in oils or petroleum fractions. The sample can either be ashed or diluted until aspirable (e.g. with 4-methylpentan-2-one), and calibration performed using special **organic standards**.

7.1.2 Agricultural samples

In soil analysis, the sample pre-treatment varies depending on whether a **total elemental** analysis or an **exchangeable cation** analysis is required. In the former, a silicate analysis method (see below) is appropriate. In the latter, the soil is shaken with an **extractant solution**, e.g. 1 M ammonium chloride, ammonium acetate or disodium EDTA. After filtration, the extractant solution is analysed. Fertilizers and crops can be treated as chemical and food samples, respectively.

AN INTRODUCTION TO ATOMIC ABSORPTION SPECTROSCOPY

Table 7.1 Detection limits for flame atomic absorption

Element		Characteristic concentration† (μg ml^{-1})	Detection limit (μg ml^{-1})
Ag	Silver	0.029	0.002
Al	Aluminium	0.75	0.018
As	Arsenic	0.92 (a)	0.11 (a)
		0.6 (b)	0.26 (b)
			0.002 (c)
Au	Gold	0.11	0.009
B	Boron	8.4	2.0
Ba	Barium	0.20	0.020
Be	Beryllium	0.016	0.0007
Bi	Bismuth	0.20	0.046
Ca	Calcium	0.013	0.002
Cd	Cadmium	0.011	0.0007
Co	Cobalt	0.053	0.007
Cr	Chromium	0.055	0.005
Cs	Caesium	0.040	0.004 (a)
Cu	Copper	0.040	0.002
Dy	Dysprosium	0.67	0.028
Er	Erbium	0.46	0.026
Eu	Europium	0.34	0.014
Fe	Iron	0.045	0.006
Ga	Gallium	0.72 (a)	0.038 (d)
Gd	Gadolinium	19	1.1
Ge	Germanium	1.3	0.11
Hf	Hafnium	10	1.4
Hg	Mercury	2.2	0.16
		0.0001 (c)	0.00004 (c)
Ho	Holmium	0.76	0.035
In	Indium	0.17	0.038
Ir	Iridium	2.3	0.36
K	Potassium	0.009	0.002
La	Lanthanum	48	2.1
Li	Lithium	0.017	0.0015
Lu	Lutetium	7.9	0.30
Mg	Magnesium	0.003	0.0002
Mn	Manganese	0.021	0.002
Mo	Molybdenum	0.28	0.030
Na	Sodium	0.003	0.0002
Nb	Niobium	19	2.9
Nd	Neodymium	6.3	1.1
Ni	Nickel	0.050	0.008
Os	Osmium	1.2	0.12
Pb	Lead	0.11	0.015
Pd	Palladium	0.092	0.016
Pr	Praseodymium	18	8.3
Pt	Platinum	1.2	0.090
Rb	Rubidium	0.030	0.002
Re	Rhenium	9.5	0.85
Rh	Rhodium	0.12	0.005
Ru	Ruthenium	0.72	0.087
Sb	Antimony	0.29	0.041
Sc	Scandium	0.27	0.025
Se	Selenium	0.38 (a)	0.25 (a)
			0.04 (e)
			0.002 (c)
Si	Silicon	1.5	0.20
Sm	Samarium	6.6	0.75
Sn	Tin	0.39	0.031
Sr	Strontium	0.041	0.002
Ta	Tantalum	11	1.8
Tb	Terbium	7.9	0.5
Te	Tellurium	0.26	0.035

THE PRACTICE OF FLAME SPECTROSCOPY

Table 7.1—continued

Element		Characteristic concentration† ($\mu g\ ml^{-1}$)	Detection limit ($\mu g\ ml^{-1}$)
Ti	Titanium	1.4	0.05
Tl	Thallium	0.28	0.013
Tm	Thulium	0.27	0.014
U	Uranium	113	59
V	Vanadium	0.75	0.05
W	Tungsten	5.8	0.52
Y	Yttrium	2.3	0.11
Yb	Ytterbium	0.073	0.0021
Zn	Zinc	0.009	0.001
Zr	Zirconium	9.1	1.0

† Characteristic concentrations and detection limits quoted were all measured at the most sensitive line. Hollow cathode lamps were used throughout together with an R446 photomultiplier. Aqueous solutions were employed for all elements and the fuel/support gas combinations were those recommended for general analytical use, except for the special cases noted. From 'Data for Varian Model AA-6 Spectrometer', supplied by Varian Techtron.

(a) Air–acetylene
(b) Air–hydrogen
(c) Vapour generation
(d) Nitrous oxide–acetylene
(e) Nitrogen–hydrogen–entrained air

7.1.3 Waters and effluents

Where the analyte is present in sufficient concentration, it may be determined **directly.** Otherwise, it may need to be **concentrated** by evaporation before determination. In either case, releasing agents and ionization buffers may be required. Electrothermal atomization (see Chapter 11) now offers a method for determining the ultra-trace metals, but if a flame is to be used for metals present in such low concentrations **solvent extraction** may be employed.

If information on **total metals** is required, the sample must be acidified before analysis. If information on **dissolved metals** only is required, the sample may be filtered (using a specified pore size) before analysis. Losses may occur, however, by adsorption during filtration.

7.1.4 Geochemical and mineralogical samples

Silicate analysis is not without problems. If measurement of silicon is not required, it may be volatilized off as silicon tetrafluoride, using hydrofluoric acid, although some calcium may be lost as calcium fluoride. Alternatively, sodium carbonate–boric acid **fusions** may be employed. Where possible, final solutions are made up in hydrochloric acid, and lanthanum is added as a buffer and releasing agent.

7.1.5 Metals

AAS is a very useful method for **calibrating** samples, even for laboratories equipped with direct-reading optical emission spectrometers and X-ray fluorescence spectrometers. Where possible, hydrochloric acid–nitric acid is used to dissolve the sample. The standards may be prepared by dissolving the trace metal in an appropriate solution of the **matrix metal** (e.g. iron(III) chloride solution for steel). If possible, 1–2% (w/v) solutions are used. Precautions against chemical interference may be necessary in flame AAS and in flame AES, where additional spectral interferences may be encountered.

7.1.6 Solvent extraction of trace metals

Solvent extraction is a particularly attractive **separation technique** for flame spectroscopy. The specificity of AAS means that only rarely is it necessary to separate an analyte from a matrix. (In AES, spectral interferences, such as by iron in steel samples, often make this more desirable.) The technique can be used to **pre-concentrate** the sample. If 100 ml of sample can be extracted using 10 ml of organic solvent, a **concentration factor** of 10 will be achieved. If larger amounts of samples (e.g. waters and effluents) are available, concentration factors up to 100 may be achieved. Additionally, a further gain in sensitivity of a factor of 3–5 will be achieved with most organic solvents, as markedly **improved nebulization** efficiencies can be obtained (see section 2.4).

A popular system uses **chelation** by ammonium pyrrolidine dithiocarbamate (the non-specificity of this reagent being an advantage for multi-element separations) followed by extraction into 4-methylpentan-2-one (methyl isobutyl ketone). This is often referred to as the **APDC–MIBK method**. Careful **control of pH** is necessary and care should be taken to use fresh chelating agent, as solutions of APDC rapidly deteriorate. Many other systems have also been proposed, and Cresser (see Appendix C) has gathered these together into a valuable book.

Q. What are the advantages of the method of standard additions?

Q. Outline methods for the determination of (i) Ca in serum; (ii) Ag in silicate rock; (iii) Mn in steel.

Q. What are the advantages of solvent extraction?

7.2 COMPARISON OF THE ANALYTICAL UTILITY OF ATOMIC EMISSION, ATOMIC ABSORPTION AND ATOMIC FLUORESCENCE SPECTROSCOPY

Increasingly, attention is being drawn to the need to regard all three techniques as **complementary**. While at present each technique has a group of elements for which it provides the most sensitive method of analysis, there are many elements for which two or even all three methods are equally sensitive. In different situations, the instrumental advantages or disadvantages of one technique may favour its use as opposed to another. Thus, it is useful to make a **summary** of some points.

(1) It has frequently been shown that the probability of the superimposition of resonance lines of different elements is extremely small, even compared with the probability that an emission line will be superimposed on the resonance line in emission methods. Even when workers have drawn attention to the occurrence of **spectral interferences** in AAS, it has been pointed out that such interferences are easily overcome by use of another resonance line (see section 6.4.1).

(2) In atomic emission, the intensity of radiation measured depends on the population of the excited level. **Small variations in temperature** (e.g. caused by fluctuations in the fuel flow rate) have a great effect on the population of excited atoms, as determined by the Boltzmann distribution (Eqn 3.2), section 3.1), and hence on the analytical signal. Indeed, the analytical signal is critically affected in emission by any interelement effect which reduces the excited population. In atomic absorption, it is the number of atoms in the unexcited state that is important, and this **ground state population** is not greatly altered by temperature variations. It is this which gives atomic absorption its main analytical advantage. As atomic fluorescence depends on the re-emission of absorbed radiation, it too is related to the ground state population and to a great extent shares this advantage of atomic absorption, although certain flame species which quench fluorescence may contribute to a loss in sensitivity. It must also be noted that in AFS, the analytical signal is critically affected by **variations in source intensity**, whereas AAS is independent of long-term source drifts since it is a ratio technique.

(3) Several workers have published studies comparing the theoretically obtainable signal strengths and **detection limits** of atomic emission, atomic absorption and atomic fluorescence spectrometry. Particular attention is drawn to the work of Alkemade (*Appl. Optics* **7**, 1261 (1968), in which equivalent noise levels are assumed in each of the techniques, and Winefordner and co-workers (*Anal. Chem.* **39**, 436 (1967), who give a complex treatment in which the theoretical noise levels are first

evaluated. Only a few points from the extensive and sometimes controversial literature will be noted here.

(a) **Atomic emission** is most sensitive at **high wavelengths** (i.e. >350 nm). This is because shorter wavelengths correspond to higher energy transitions, and even high-temperature flames, such as the nitrous oxide–acetylene flame, do not possess enough thermal energy to produce sufficient atomic populations of higher excited states.

(b) Alkemade has derived a ratio for the relative intensities of atomic absorption (ΔI) and atomic emission (I_{em}) neglecting any differences in noise levels in the two methods:

$$\frac{\Delta I}{I_{em}} = \frac{J_s}{J_{pl}(T_F)} \frac{\Delta \lambda_s}{\Delta \lambda_F} \qquad (7.1)$$

where J_s is the spectral radiance of the source (atomic absorption), $J_{pl}(T_F)$ is the spectral radiance of a black body at the atomic cell (in this case, a flame) temperature, T_F, at the wavelength λ of the line centre (given by the Planck radiation law), and $\Delta \lambda_s$ and $\Delta \lambda_F$ are the spectral width of the source and emission/absorption line, respectively. Thus, if $\Delta \lambda_s \approx \Delta \lambda_F$, atomic absorption is only more sensitive if the **spectral radiance** of the source exceeds that of a **black body** at the temperature of the flame and at the wavelength of the analysis line. This condition is met in a number of situations, e.g. for the ultraviolet region of the spectrum, in a hollow-cathode lamp and in gas discharge tubes.

(c) Especially in the case of relatively high-energy states (i.e. from which transitions give rise to resonance lines in the spectral region below 300 nm), it has been shown that, under suitable conditions (a source with high integrated radiance at the wavelength of interest), the excited state population in atomic fluorescence can considerably exceed the thermal value, i.e. the effective population for atomic emission.

Thus, if the same atomic resonance line is measured by all three methods using the same instrumental system, it can be shown that atomic emission should result in **lower limits of detection** than atomic fluorescence or atomic absorption for resonance lines above about 400 nm. Between 300 and 400 nm, all three methods should give similar limits, but below 300 nm, atomic absorption and atomic fluorescence should provide lower limits of detection. Published experimental results tend to confirm this generalization.

THE PRACTICE OF FLAME SPECTROSCOPY

(4) In absorption methods, the **ratio** of the unabsorbed signal to the absorbed signal is measured. This is an advantage of the atomic absorption technique, in that it **removes** the possibility of **systematic errors** arising from small shifts in the monochromator setting, or small variations in the sensitivity of the detection system between optimization and the completion of the analysis. However, it is experimentally difficult to measure a **small difference in two large quantities**. This places a practical limit on the sensitivity of absorption techniques, whereas the sensitivity of atomic emission and atomic fluorescence are not limited in this way. The sensitivity of atomic emission can be increased by the use of atom cells with **higher temperatures** (e.g. inductively coupled plasmas), and that of atomic fluorescence, by the use of **brighter sources**.

(5) As AES requires no spectral source, it is more **easily adapted** to non-standard and qualitative determinations. However, it is more prone to **spectral interference**.

(6) Compared with the two emission methods, AAS requires **less operator skill** to give results of high reproducibility and good sensitivity (another advantage of the ratio method).

(7) AES and AFS can be readily adapted to **simultaneous multi-element** analysis, whereas AAS, because of the necessary geometry, is far less readily adaptable.

(8) Generally, the dynamic **working range** (region of linear calibration curves) is in the order AFS > AES > AAS.

(9) AAS is readily applicable to the **widest range** of elements, the sensitivity obtainable is independent of the source intensity or operator skill, and it is applicable with minimal training.

Q. Which technique suffers most from spectral interferences and why?

Q. Cadmium (primary resonance line 228.8 nm) can be determined more sensitively by AAS than by AES, whereas for sodium (primary resonance line 589 nm) the reverse is true. Why?

Q. Give some reasons why, of the three techniques, AAS is generally regarded as the simplest to operate.

Q. Why is AAS least suitable for simultaneous multi-element analysis?

Q. Which technique would you recommend for (i) routine determination of Zn in effluent; (ii) non-routine determination of low level Li?

8 MODIFICATIONS TO FLAME SPECTROSCOPY

8.1 THE LIMITATIONS OF CONVENTIONAL FLAME CELLS

Some fundamental and possible **practical disadvantages** of the use of flames for analytical atomic spectroscopy are given below and in section 8.3. The disadvantages can be dealt with in two lists. Those in the first may be ameliorated by modifying the flame. The second list (in section 8.3) outlines some more fundamental restrictions.

(1) The conventional indirect nebulizer and flame systems require relatively **large volumes of solution** to operate. The nebulizer is designed to supply a fine aerosol of solution to the burner at a uniform rate. In these systems, large sample droplets condense on the walls of the spray chamber. Hence, only about 10% of solution uptake is delivered to the flame. Attempts to improve the efficiency of premix nebulizers by using **ultrasonic nebulization**, **heated spray chambers** or **hot gases** have been reported, but generally the apparatus used tended to lack simplicity.

(2) Atom concentrations in flames are limited by the **dilution effects** of the relatively high flow-rate of unburnt gas used to support the flame, and the **flame gas expansion** which occurs on combustion. It is estimated that the atoms **spend 10^{-4} s** in the analysis volume, yet it is practically difficult to record the equilibrium signal from a sample solution in less than 10 s.

(3) Solutions with high sample matrix or analyte concentrations may undergo incomplete solute vaporization and gaseous dissociation because of the **short transit time** in the flame cell. Some elements are prone to **compound formation**. Although this can be overcome by use of the hot, reducing, fuel-rich nitrous oxide–acetylene flame, sufficient partial pressures of oxygen-containing species may be available to reduce the concentration of some elements which form stable monoxides (e.g. Zr, Hf, Si, B). Some elements are strongly **ionized** in flames (e.g. Na, K, Rb,

MODIFICATIONS TO FLAME SPECTROSCOPY

Cs), and a more-readily ionized element must be added to suppress this ionization if the atomic population is not to be seriously reduced.

(4) Although in many ways **solutions** are ideal analytical matrices, in certain situations the analysis of **solid samples** may be preferred. It is difficult to nebulize viscous oils, and certain other common organic solvents may extinguish the flame when sprayed.

In AAS, the only remaining practical way to lower limits of detection seems to be to increase the efficiency of the atom cell. This has led to considerable interest in new and modified flame cells.

Q. In conventional flame spectroscopy, where does most of the solution go?

Q. Why do atoms spend such a short time in the analysis volume?

Q. What features should any 'new' atom cell for AAS have?

8.2 MODIFIED FLAME CELLS

To overcome the above limitations, a number of modifications to the flame cell have been proposed.

8.2.1 Pulse nebulization

This procedure permits the use of **smaller samples** (25–200 mm^3) and **higher concentrations** (e.g. 10% (w/v) steel) than in normal nebulization. A **cup**, made of Teflon or another suitable plastic, is attached to the nebulizer tube, and the sample pipetted into the cup using a **micro-pipette**. The sample is totally consumed and a **peak signal** is observed. We have used the technique in steel analysis and in a biological study where only a very small amount of sample was available. The technique also enjoys a number of alternative names: discrete sample nebulization; gulp sampling; direct-injection cup nebulization; Hoescht cup nebulization.

8.2.2 Branched uptake capillaries

If the capillary uptake to the nebulizer is branched (using, for example, a **T-piece** from an auto-analyzer), **two capillaries** may be connected. Thus, a buffer or ionization suppressor may be added without time-consuming solution preparation. Similarly, standard additions may be performed in this way. We have shown that a branched capillary may be used to enable calibration with aqueous standards when analysing organic extracts.

8.2.3 Kahn sampling boat

Kahn *et al.* (*At. Absorpt. Newsl.* **7**, 35 (1968)) developed a system where nebulizer inefficiency was avoided by using a **tantalum** sampling boat, from which the sample was evaporated as it was pushed into the flame. Obviously, the technique is only applicable to the more-easily atomized elements, but it gave a useful improvement in sensitivity. Unfortunately, the technique is prone to poor reproducibility.

8.2.4 Delves sampling cup

Delves (*Analyst* **95**, 431 (1970)) successfully modified the above technique and applied it to the determination of **lead in micro-samples** of whole blood. The Delves cup has particular application for the determination of lead in a variety of samples. For example, we have recently used the technique to determine lead in road-side grass, without the need for digestion.

The tantalum boat is replaced with a smaller (10 mm outer diameter, 5 mm deep, 0.15 mm metal foil) and more-easily positioned **nickel microcrucible** (or more recently, stainless-steel crucible). The crucible may also be used during preliminary chemical treatment (e.g. addition of hydrogen peroxide to destroy organic matter). It is mounted into a device which enables it to be pushed close to the flame, to allow charring, and then into the flame, to allow atomization. A **nickel absorption tube** is mounted in the flame, and the atoms enter the tube through a hole half-way along its length. Light passes through the tube, thus improving reproducibility by lengthening and defining the **residence** of the atoms in the flame. More recently, **silica** absorption tubes have been used with a considerable increase in lifetime.

The whole device is shown in Fig. 8.1, which was taken from Delves' original paper.

Q. What devices can be used when only small volumes of sample are available?

Q. Give two ways in which sample pre-treatment may be minimized.

Q. What are the primary advantages and disadvantages of the Delves cup technique?

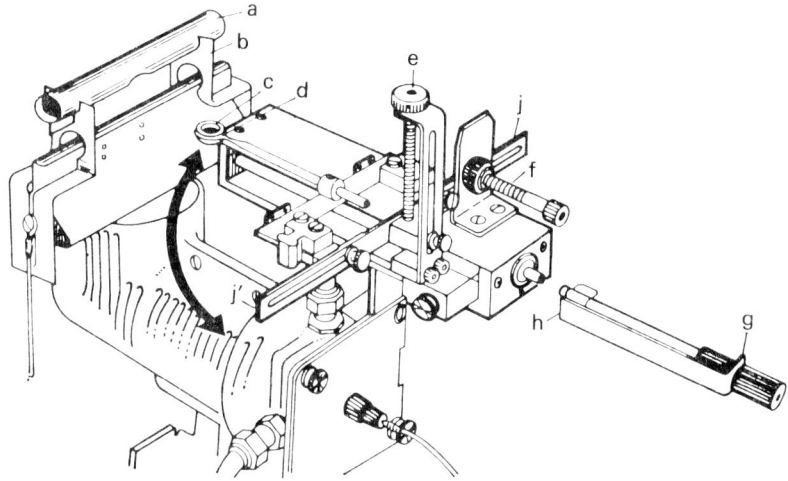

Fig. 8.1 Delves microsampling cup. (a) Nickel absorption tube (100 mm long × 12.5 mm o.d.), made from 0.006-inch Ni foil; (b) nickel supports for (a), made from 0.020-inch Ni foil; (c) nickel crucible (10 mm o.d. × 5 mm tall), made from 0.006-inch Ni foil; (d) platinum-wire holder (0.5 mm diameter) for (c), sealed in a glass tube; (e) vertical adjustment screw; (f) horizontal angular adjustment screw; (g) adjustment screw for slide stop (h); (j and j') rotating limb.

8.3 FURTHER ARGUMENTS ABOUT FLAME ATOM CELLS

It is important to remember how useful flames are. They are still used for the **majority of analytical determinations**, and here we list some of their advantages.

(1) Flames have a **long history** of use and much is known about their behaviour.
(2) Most elements can be easily **atomized** by the appropriate flame, most commonly air–acetylene or nitrous oxide–acetylene.
(3) Flame cells are easily **optimized**, and there is a choice of flame types.
(4) A **steady** (non-transient) signal is obtained.
(5) **Burners** are easily fabricated and durable, and adaptable to give either long or short path lengths.
(6) Flames are **simple to use**, reliable and relatively free from memory effects.
(7) The **signal-to-background** and **signal-to-noise** ratios obtainable for most elements allow **good sensitivity and precision** (0.4–2% r.s.d.) to be achieved in a wide range of practical analyses at different wavelengths between 200 and 800 nm.
(8) **Automatic sampling** devices can conveniently be used in conjunction with flames.

(9) A **rapid rate** of analysis is achievable with a flame system.

Some disadvantages of flames have already been noted in section 8.1, and we must accept that, while the modifications to flames may overcome some of these problems, there are **further disadvantages** of flame cells that these modifications, which are mainly applicable to volatile elements, do not overcome.

(1) Flames exhibit a **background radiation** and absorption which consist of **banded and continuous** spectra. The banded spectra arise from the excited molecules and radicals in the flame gases, and the continuous spectra from the dissociation, ionization and recombination of these species. Care must be taken that the line of analytical interest is not in a region of high flame background. Problems associated with flame background can be reduced by separating the secondary diffusion flame from the primary reaction zone, and viewing the low-background interconal zone (see section 2.6). In flames, considerable reduction in radiation intensity at wavelengths below 200 nm is caused by absorption of radiation by flame gas products, whereas flames also produce appreciable thermal excitation of elements having resonance lines greater than 300 nm (the effects of such emission signals can be minimized by synchronous modulation and amplification). Thus, flames possess considerable disadvantages for the atomic absorption or fluorescence of elements whose strongest resonance line is below 200 nm or above 300 nm.

(2) In atomic fluorescence, the flame which produces the greatest freedom from inter-element effects may also produce **lower fluorescence efficiency** than cooler, hydrogen-based flames, because of quenching of radiation-excited analyte atoms by the molecules in the flame gas.

(3) Particulate matter in the light path may **scatter** radiation.

(4) In certain situations, flames are **inconvenient**. Either the flame or the associated high-pressure cyclinders may present a **hazard**. It is not advisable to spray radioactive solutions into a flame, nor to leave flames unattended.

(5) Flame gases can be comparatively expensive, and **extraction systems** are necessary. **Safety regulations** concerning the use of acetylene and particularly the undesirability of storing this gas in the laboratory may involve considerable cost and laboratory reorganization.

(6) Explosion **hazards** are always present with flames of high burning velocity and flame products may be toxic. **Hot flames**, besides often being unpleasant to use, can affect the auxiliary equipment, e.g. cause monochromator drift.

(7) **Precise control** over species in the flame is **limited**. The chemical composition of the flame or its temperature can be only controlled over a very small range and generally only by altering other parameters. It is difficult to envisage the use of flame cells in absolute analytical work.

MODIFICATIONS TO FLAME SPECTROSCOPY

Q. List some advantages of flames as atomizers and compare them with the list above.

Q. What problems does background radiation introduce into AAS?

Q. List some of the hazards associated with using flames.

9 MERCURY DETERMINATION

9.1 THE REDUCTION–AERATION METHOD

Mercury is worthy of particular mention because (i) it has attracted very considerable attention in the light of its **environmental importance** and (ii) it is the only element with an appreciable atomic **vapour pressure** at **room temperature**.

The 253.7 nm line is usually used for mercury atomic absorption, but this transition is **spin-forbidden** and relatively **insensitive**. The 184.9 nm line is potentially 20–40 times **more sensitive**, but at this wavelength most flame gases and the atmosphere **absorb strongly**. Thus, flame methods for mercury are not noted for their sensitivity (typical flame detection limits are in the range 1–0.1 ppm). If **chemical reduction** is employed, mercury can be brought into the vapour phase without the need to use a flame, and detection limits can be dramatically **improved**.

Four main methods have been used to bring mercury into the vapour phase.

(i) **Reduction–aeration**: mercury in aqueous solution is treated with a reducing agent and then swept out of solution by bubbling a gas into it.
(ii) **Heating**: the sample is pyrolysed or combusted.
(iii) **Electrolytic amalgamation**: mercury is plated on to a copper cathode during electrolysis. The cathode is then heated as in (ii), to release the mercury.
(iv) **Direct amalgamation**: mercury is collected on a silver or gold wire, from which it is then released by heating. This method may be employed, with (i) and (ii), as a concentration method.

A **review** by Ure (*Anal. Chim. Acta* **76**, 1 (1975)) remains the most comprehensive survey of these various approaches, and notes over 400 applications.

The reduction–aeration method is by far the most popular. The first report of this method was by Hatch and Ott (*Anal. Chem.* **40**, 2085 (1968)), who used **tin(II)** sulphate solution as the **reducing agent**. Tin(II) chloride (stannous chloride) is presently the most widely used reductant. Some workers favour **sodium borohydride**, but this can pose safety problems, and the evolution of large amounts of hydrogen increases quenching if AFS is employed. In fact, the mercury will **equipartition** between the aqueous sample and the head-space. Therefore, bubbling is not essential, but it does improve the speed of analyis.

The method is applicable to a **wide range** of samples. Most samples require prior oxidation to ensure that all the mercury is in an inorganic form. Excess oxidizing agents (e.g. permanganate) can be destroyed by the use of a **mild reducing agent** (e.g. hydroxylamine hydrochloride) prior to reduction–aeration. It is even reported that different forms of mercury can be determined separately by this method. Inorganic mercury is determined by complexing the mercury with **cysteine** in acidic solution, followed by reduction–aeration in alkaline solution. If a cadmium or copper salt is also added, mercury is liberated from both organic and inorganic mercury compounds. The organic mercury content can be obtained by difference.

Q. Why are special methods for mercury (i) desirable; (ii) feasible?

Q. What are the pertinent chemical equations for the reduction–aeration method?

Q. How may organically and inorganically bound mercury be determined in a given sample?

9.2 COLD-VAPOUR ATOMIC ABSORPTION SPECTROSCOPY

The concentration of the resultant mercury vapour (often referred to as a cold-vapour) can be determined by AAS or AFS.

A system for cold-vapour AAS is shown in Fig. 9.1. The evolved mercury vapour is passed to a long path-length **absorption cell**, usually constructed of Pyrex glass tubing with silica end windows. A **transient** absorption peak is observed. In some systems, a **recirculating pump** is used to cycle the mercury vapour around the system and achieve a **steady reading**.

Such arrangements as Fig. 9.1 are exceedingly simple and can be built 'in house'. A hollow cathode lamp is used as source and the cell can be taped to the top of the burner. Air from a small compressor can be used for aeration. As an **enclosed cell** is used, problems can arise from water vapour condensation and spray. The cell can be heated to minimize such problems. Spurious absorption signals can often be seen and the use of **background correction** is to be preferred. Detection limits of about **1 ng** are achievable. Certain elements, particularly silver, gold, palladium and platinum, are reported to **interfere** in the reduction stage.

AN INTRODUCTION TO ATOMIC ABSORPTION SPECTROSCOPY

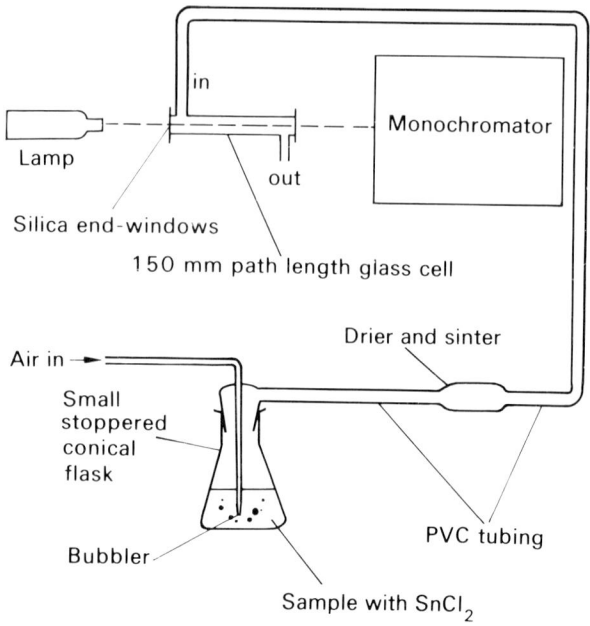

Fig. 9.1 Cold vapour determination of mercury by atomic absorption spectroscopy.

Q. List the advantages and disadvantages of cold-vapour AAS for mercury.

9.3 COLD-VAPOUR ATOMIC FLUORESCENCE SPECTROSCOPY

The need for an enclosed cell can be overcome by using AFS. In a system we have developed (Fig. 9.2), a laminar sheath of argon encloses the mercury plume. **Argon** must be used for the flushing as it minimizes quenching of the excited mercury species. For the reasons given in section 9.1, tin(II) chloride is used as reductant. The **sheathing gas flow** is maintained by passing argon through a concentric bundle of glass capillaries glued to the sample delivery tube.

The major problem with the AFS technique tends to be with **sources**. Electrodeless discharge lamps are very intense, but often lack stability. Low-pressure vapour discharge lamps offer better stability, but poorer intensity. Detection limits in the **picogram** range are achievable both with electrodeless discharge lamp sources and vapour discharge lamps.

MERCURY DETERMINATION

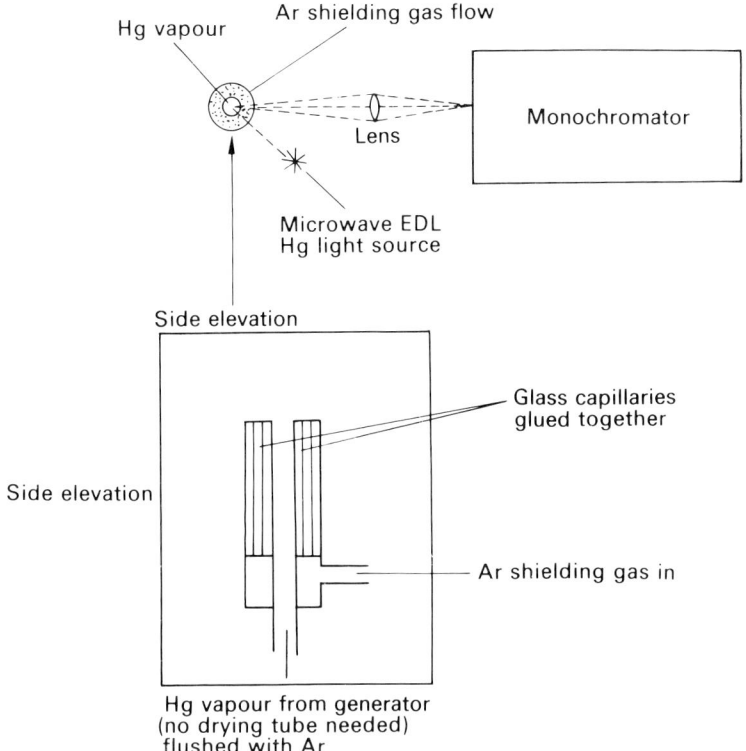

Fig. 9.2 Atomic fluorescence spectroscopic system for the cold vapour determination of mercury.

Q. List the advantages and disadvantages of cold-vapour AFS for mercury.

10 HYDRIDE GENERATION

10.1 CHEMISTRY

A number of elements are now routinely determined by generating their **covalent gaseous hydrides** and atomizing these, either in a flame or an electrically heated tube. Initially, the method was applied to **arsenic** (Manning, *At. Absorpt. Newsl.* **10**, 86 (1971)) **and selenium** (*ibid*, p. 123), both of which pose difficulties in AAS because of their low-wavelength, primary-resonance lines. The hydrides were generated by **zinc–hydrochloric acid reduction** and collected in a balloon, before expulsion into an **argon–hydrogen diffusion flame**.

A better reducing agent is **sodium borohydride**, because the hydrides are formed more **rapidly** and a collection reservoir is not needed. A 1% (w/v) aqueous solution is usually sufficient, provided there is good stirring of the acidified sample. This reagent can be used to generate the **hydrides of antimony, arsenic, bismuth, germanium, lead, selenium, tellurium and tin**. The replacement of inefficient nebulization by **gaseous sample transport** improves the detection limit of all the listed elements except lead (the improvement for germanium is not dramatic), where the limiting factor is presumably the difficulty in forming plumbane. If the lead is oxidized prior to reduction, improvements in detectability are also observed with this element (presumably an unstable intermediate lead(IV) state is involved).

Arsenic and antimony also give responses which **vary according to valence state**, the +5 state giving poorer responses than the +3 state. These elements are therefore reduced in iodide solution prior to hydride generation. Similarly, selenium and tellurium should be in the +4, not the +6, state.

Considerable **inter-element interference** effects have been reported on the actual hydride formation step. Elements easily reduced by sodium borohydride (e.g. silver, gold, copper, nickel) give rise to the greatest suppressions. Interferences in the actual atom cell have not been documented.

AAS is usually used to complete the analysis, and instrumentation for this is described in the next section. As most of the hydride-forming elements are particularly favorable for **AFS**, this can also be used, with resultant improvements in detection limits.

HYDRIDE GENERATION

Q. What feature of hydride generation methods gives rise to the improvement in sensitivity associated with them?

Q. Which elements can usefully be determined via hydride generation?

Q. Which elements are common interferents in hydride generation methods?

10.2 INSTRUMENTATION

In most arrangements, the acidified sample (1 cm^3) is injected into a stirred glass cell containing 1% (w/v) aqueous sodium borohydride (2 cm^3). The contents are stirred and the **liberated hydrides flushed** with nitrogen into either:

(i) a **flame**, often an argon–hydrogen diffusion flame;
(ii) a narrow-bore **silica tube** mounted over an air–acetylene flame (in one design, there is a transverse flow of nitrogen at the ends of the tube to ensure that liberated hydrogen does not burn in the light path—some results from such a system are shown in Table 10.1);
(iii) a narrow-bore **silica tube, electrically heated** by means of a winding of suitable resistance wire.

The use of narrow-bore tubing results in much improved limits of detection by **limiting the dilution** of the hydrides. Using arrangements (ii) or (iii), background correction is usually unnecessary, provided hydrogen is not allowed to burn in the optical axis.

Table 10.1

Element	Wavelength (nm)	Reductant	Characteristic concentration (μg cm^{-3})	Detection limit (μg cm^{-3})
As	193.7	NaBH$_4$	0.00052	0.0008
		Zn–HCl	0.001	0.0015
Bi	223.1	NaBH$_4$	0.00043	0.0002
Ge	265.1	NaBH$_4$	1.0	0.5
Pb	283.3	NaBH$_4$	0.08	0.1
Sb	217.6	NaBH$_4$	0.00061	0.0005
Se	196.1	NaBH$_4$	0.0021	0.0018
Sn	224.6	NaBH$_4$	0.00044	0.0005
Te	214.3	NaBH$_4$	0.002	0.0015

From Thompson and Thomerson, *Analyst* **99**, 595 (1974). Silica tube in air–acetylene flame; nitrogen flushed.

Q. Why is it preferable to atomize the hydrides in a narrow-bore tube?

11 ELECTROTHERMAL ATOMIZATION

11.1 HISTORICAL DEVELOPMENT

The disadvantages of flames have been outlined above. Not unnaturally, the attention of spectroscopists and analytical chemists has been drawn to the use of **other devices for atomization**. The use of arcs, sparks, the DC arc plasma, the r.f. inductively coupled plasma, the capacitatively coupled microwave plasma, the microwave induced plasma, the hollow cathode sputtering cell, the glow discharge lamp, lasers, flash lamps and induction furnaces for atomization will not be covered in detail in this book, although several of these techniques have been outlined elsewhere (section 3.3). This discussion will be confined to **atomization by thermal means** following resistive heating, so called **electrothermal atomization**.

In 1905 and 1908, **King**, generally regarded as the first worker in this field, reported on fundamental spectral studies using an electrically heated tubular furnace. The classical work in analytical chemistry is that of **L'vov**, who began to publish his results in 1959. Here the sample was applied to the tip of a **carbon electrode** which was introduced into a cylindrical **heated furnace** through a transverse aperture at the centre of the tube. At first, the sample was preheated using a powerful DC arc (arced to the electrode with the sample in position). Later this arrangement was replaced by simpler **resistive heating of the electrode** (Fig. 11.1). The graphite tube was 30–50 mm in length, with an internal diameter of 2.5–5.0 mm and an external diameter of 6.0 mm. At first, the cylinders were lined with tungsten or tantalum foil to retard vapour diffusion, but later this was changed to a coating of **pyrolytic graphite**. The tube did not act as an atomizing furnace, merely as an **atom cell** hindering the loss of atoms. The sample electrode was responsible for atomization. Both the tube and the electrode were heated via **step-down transformers** of 4 kW (at 10 V) and 1 kW (at 15 V), respectively. The atomizer was placed within a chamber filled with **argon or nitrogen** and light passed down the tube for AAS measurements. L'vov's apparatus offers the best absolute sensitivities for AAS yet obtained, but it has been criti-

ELECTROTHERMAL ATOMIZATION

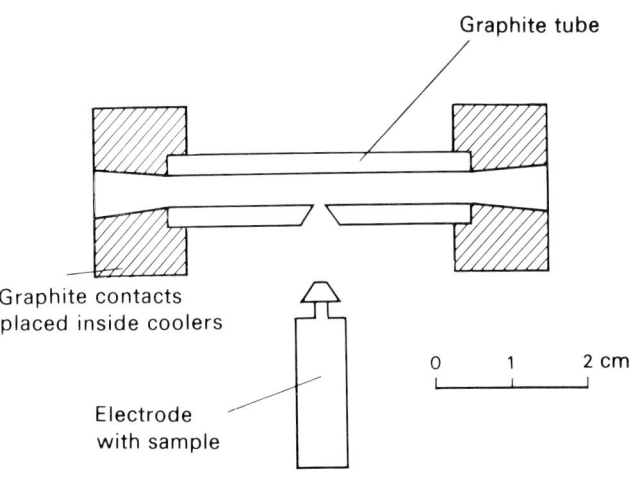

Fig. 11.1   Graphite furnace of L'vov.

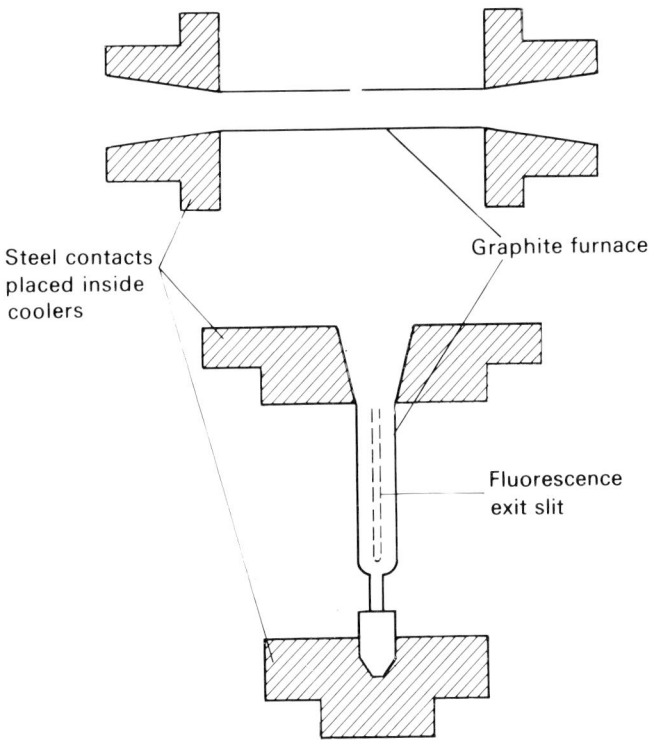

Fig. 11.2 Massmann furnaces for (top) atomic absorption and (bottom) atomic fluorescence.

cised for being too cumbersome. Current commercial atomizers are based on two simpler designs.

In 1967, **Massmann** described a heated **graphite furnace** in which no auxiliary electrode was used, i.e. the graphite tube was both the resistance element and the furnace. The sample was micropipetted directly into a 55 mm long, 6.5 mm internal diameter, 1.5 mm wall thickness tube via a small, 2 mm diameter orifice. The absorption tube device and a graphite cup for AFS are shown in Fig. 11.2. Using a power supply of 400 A at 10 V, the furnace could be heated up to 2600 °C in a few seconds. Typical solution volumes of 5–200 mm^3 were used.

In the **West rod** atomizer, first reported in 1969, no tube was used; the sample was applied directly to an **electrically heated filament**. The graphite filament (2 mm diameter, 40 mm long), supported by water cooled electrodes, could be heated to 2000–3000 °C within 5 s by the passage of a current of about 70 A at up to 12 V. Small liquid samples (1–5 mm^3) were pipetted on to a depression on the rod. While the original filament was enclosed in an **inert gas** purged chamber, it was later found simpler to **shield** the graphite from the air by a simple flow of shielding gas around the filament. The apparatus is shown in Fig. 11.3. The greatest sensitivity and free-

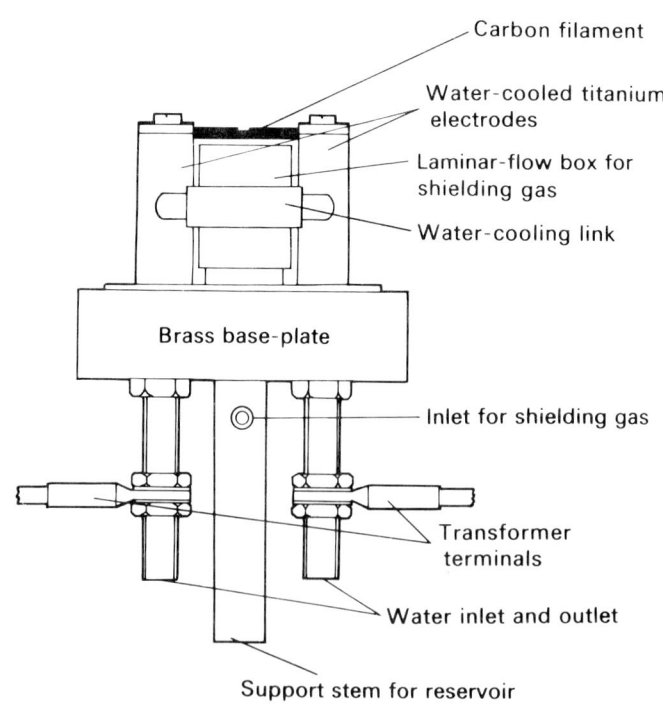

Fig. 11.3   West rod atomizer.

dom from matrix effects was found when making observations of the atom cloud immediately above the filament. The device is equally applicable for AAS (Ebdon et al., Anal. Chim. Acta **58**, 39 (1971)) and AFS (Ebdon et al., Talanta **19**, 1301 (1972)). By **drilling holes** in the filament, a compromise between the extended thermal contact of the furnace and the simplicity of the filament (or rod) has been reported.

Kirkbright has **reviewed** the history of electrothermal atomization both in his book (see Appendix C) and in Analyst (**96**, 609 (1971)). Fuller has published a monograph on electrothermal atomization for AAS (see Appendix C).

Q. What is the material from which all three of the above atomizers are made, and how is it heated?

Q. From what part of the atomizer is the sample atomized in (i) the L'vov cuvette, (ii) the Massmann furnace and (iii) the West rod?

Q. Bearing in mind the above, in what way can the designs of Massmann and West be regarded as simplifications of L'vov original design?

11.2 HEATED GRAPHITE FURNACE ATOMIZERS

A typical commercial furnace atomizer based mainly on the work of **Massmann** is shown in Fig. 11.4.1. Typically, the **tubes** are 25–50 mm long and 5–10 mm in diameter. The tube may be turned down at the centre to increase the temperature at that point, or the whole tube may taper towards the centre ('profiling') to shape the tube to the optical beam and increase the free atom density at the centre.

The two major disadvantages of graphite are its porosity and tendency to carbide formation. These may be partially overcome by coating the tube with **pyrolytic graphite** (e.g. by heating the tubes in a methane atmosphere) which is far less porous. Some workers have also inserted **other linings** (e.g. tungsten and tantalum) into the furnace, or deposited a carbide lining on the inner wall (e.g. using lanthanum salts).

The furnace is heated by **low voltage** (usually 10 V) and **high current** (up to 500 A) from a well stabilized step-down transformer. For optimum precision, the voltage should be well stabilized, often by a feed-back loop which may be temperature feed-back based (see section 11.4). A rapid rise-time of the temperature is also preferable, because of theoretical considerations of peak shapes. This has implications for power supply design and furnace design, as will be seen below. Currently, furnaces are available

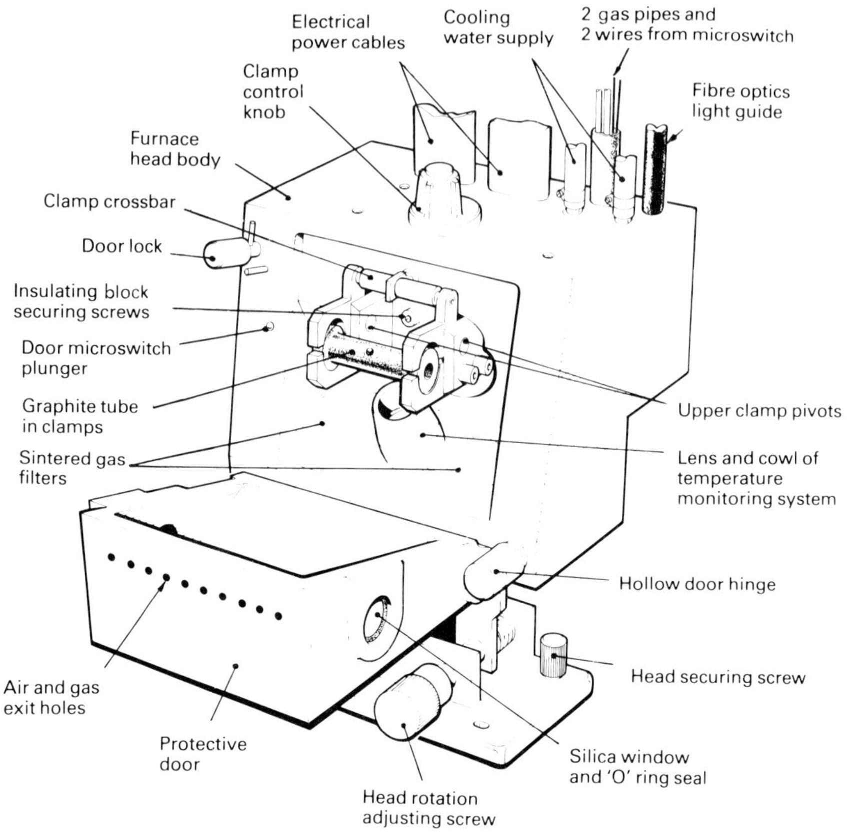

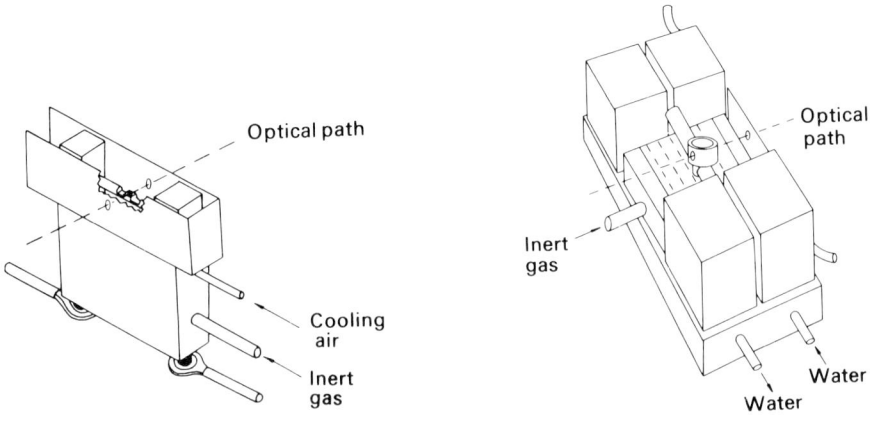

Fig. 11.4.1 Heated-graphite furnace. Section of the Pye–Unicam model HGA-2000 heated-graphite furnace.

Fig. 11.4.2 Heated-graphite filament. Schematic diagram of the Shandon Southern A3470.

Fig. 11.4.3 Heated-graphite cup atomizer. Schematic diagram of the Varian Techtron CRA 63.

which reach temperatures in excess of **3000 °C**, and temperatures of 2500 °C should be reached in less than 2 s in a well designed furnace.

The furnace is **purged** with an inert gas, usually **nitrogen or argon**. Argon, with a small addition of methane, is also used to provide continuous pyrolytic coating. There seem to be some **chemical effects** between nitrogen and certain elements (e.g. titanium, vanadium and barium), and the **rate of diffusion** of argon is less. This latter effect means that slightly larger signals are usually observed in argon. A gas flow that sweeps into the tube and out of the centre hole has been shown to reduce problems of background scatter. The flow may often be stopped during atomization to prevent dilution.

The whole atomizer may be **water cooled** to improve precision and increase the speed of analysis. The tube is positioned in place of the burner so that the light path passes through it. Liquid samples (5–100 mm^3) are placed in the furnace, via the injection hole in the centre, usually using a **micropipette** with a disposable, 'dart-like' tip. **Solid samples** may also be introduced; in some designs, this may be achieved using special graphite boats. The sample introduction step is usually the main **source of imprecision** and may also be a source of **contamination** (e.g. by zinc). Various **automatic samplers** are now available which improve precision and enable unattended operation of the furnace. These samplers are of two types; automatic injectors, and a type in which the sample is nebulized into the furnace prior to atomization.

The power supply controlling the furnace can be **programmed** so as to **dry** the sample after injection, **ash** it at an intermediate temperature (say 500 °C) and **atomize** it. The temperature and duration of each of these steps can usually be controlled over a wide range. Optimizing the operating conditions of the furnace (or **'programming the furnace'**) is a vital step in the development of analytical methods. In the drying phase, the solvent must be driven off without problems from **'spitting'**. Drying of organic solvents tends to give particular problems, and the ashing conditions are most critical. It is essential to remove organic matter by **pyrolysis** and as many volatile components of the matrix as possible, but to avoid any loss of the analyte, either as the element or as a volatile salt, such as a halide. The atomization temperature is usually chosen so as to give a **rapid peak**. It should not be so hot as to unnecessarily damage the tube or distil off involatile contaminants, nor so cool as to lose sensitivity or create memory effects (although a tube-clean, i.e. a high temperature cycle, can be included in the programme).

A spectrometer with **rapid response electronics** should be used for electrothermal atomization, as it must follow the **transient absorption event** in the tube. **Automatic simultaneous background correction** (see section 4.2.3) is virtually essential, as non-specific absorption problems in furnaces are very severe. It is important that the continuum light follows exactly the **same path** through the furnace as the radiation from the line source. The time interval between the two source pulses should be as short as possible (a chopping frequency of at least 50 Hz) because of the transient nature of the signal.

Usually, the **peak height** of the transient absorption signal is measured using either a fast-response recorder (less than 0.2 s for full-scale deflection) or a peak height mode on the instrument read-out. **Peak area** measurement often offers greater precision, and many instruments have such a facility built in. A display of the furnace peak shape on a visual display unit (or video screen) is a useful diagnostic tool now commercially available.

The majority of furnace applications concern AAS, but it is also possible to observe AES in the furnace, and this can be used analytically.

Q. What are the advantages of pyrolytic graphite coatings for furnaces?

Q. Why is argon preferred to nitrogen as the purge gas?

Q. Why are fast-response electronics and background correction essential when using furnace atomization?

11.3 FILAMENT AND MINI-FURNACE GRAPHITE ATOMIZERS

The other types of commercially based atomizers are based on the West rod. This has the advantage of **simplicity** of design, **low power** requirements and a **fast heating** rate. The disadvantage is generally regarded as the interference effects encountered because of the **rapid cooling** of the atoms once they leave the filament.

Although the original filaments were of 2 mm diameter, commercial filaments are often wider (e.g. 3.05 mm) and specially shaped to ensure **localized heating** around a central sample cavity (see Fig. 11.4.2). Sample volumes of 0.5–10 mm^3 are used and the light beam is set to **graze the surface** of the rod. The pillars of the rod may be air- or water-cooled. The power requirements are modest (1.5 kW, cf. 3.7 kW for a small furnace), but temperatures in excess of 3000 °C can be reached in 2 s. The rods can be pyrolytically coated in a similar way to tubes.

Tantalum filament atomizers have been marketed, but they are not as versatile as graphite atomizers. The metal becomes embrittled, is attacked by certain acids and may distort in shape.

A more popular modification has been to incorporate into the rod a cavity or 'mini-furnace'. At first, this was obtained by drilling a hole in the rod, injecting the sample into this cavity and passing the source light through the cavity. Latterly, this has been superseded by placing a **hollow graphite cylinder** (9 mm long, 3 mm internal diameter) between two spring-loaded

ELECTROTHERMAL ATOMIZATION

rods. This device will accept samples up to 10 mm^3 in volume (20 mm^3, with an internal thread cut into the cylinder) and achieves rapid localized heating. A moderate, 2–3 kW, power supply is used and temperatures up to 3000 °C can be reached. In the design shown in Fig. 11.4.3, a **graphite cup** is inserted instead of the cylinder. This will accept **liquid** (up to 20 mm^3) **or solid samples** and can be heated to 3000 °C within 2 s. The whole device is **inert gas shielded**, or it may be surrounded by an argon–hydrogen diffusion flame.

Filaments and cups are applicable primarily to **AAS**, but also to **AFS**. Strictures similar to those given in section 11.2 concerning **instrumentation** apply to spectrometers used with these devices.

Q. What are the advantages and disadvantages of filaments compared with furnaces?

Q. How has the original design of West been modified to minimize these disadvantages?

11.4 RECENT DEVELOPMENTS IN ELECTROTHERMAL ATOMIZER DESIGNS

With hindsight, it is now possible to identify certain **directions** in which the designs of electrothermal atomizers have logically moved. We have seen that new designs of furnaces have tended to be **smaller** than their predecessors, and that filaments have developed into **'mini-furnaces'**. There has been a **convergence** of these two designs. A fresh look has also been taken at two early designs, that of **L'vov** and that of **Woodriff**, of furnaces once dismissed as too cumbersome. The L'vov furnace has been described in section 11.1. Woodriff's furnace operated at **constant temperature** and samples were pushed into the furnace on small graphite boats.

11.4.1 Control of furnace temperature

The **faster** the rate of heating, the higher the **density** of the atoms which will be formed in the **transient** atomic cloud. This leads to improved analytical **sensitivity**, provided that the electronics can follow the more **rapid signals**. To obtain good reproducibility, it is necessary to **control the temperature** actually reached by the tube. While it is relatively simple to stabilize the **applied voltage**, variations in the **resistance** of individual tubes and the degradation of the tube as it is used mean that the temperature achieved must be **monitored**. Thus, the voltage applied should be controlled via a **feedback circuit**, linked to some method of sensing the tube temperature. This temperature may be measured using a **thermocouple**, which unfortunately may suffer from temperature lag, or by using a **light sensor**, e.g. an infrared sensor which views the tube-wall radiation via a fibre optic. In the latter case, it is clearly important that the end of the fibre optic remains clean. In a well designed modern furnace, therefore, the control settings 'dry', 'ash', 'atomize' and 'clean' should refer to **reproducible temperatures** (probably inaccurately known, but precise) rather than to different applied powers.

11.4.2 Tube dimensions

Initially, it was thought that maximum sensitivity would be obtained from larger tubes, as these would retain the atomic vapour for a longer period. Long tubes, however, require **more power** to achieve a given temperature, and **smaller tubes** can be heated more **rapidly**. Thus, in the interests of rapid heating and **simplicity** of power supply, small tubes, typically of 20–30 mm length, are preferred. The tube should not be so short that the atoms **escape** from the tube and cool too rapidly; modern tubes give residence times of about 0.5 s.

11.4.3 Isothermal operation

The early L'vov and Woodriff designs both shared the advantage that the sample was atomized into a **hot environment** (so-called isothermal atomization). Conversely, in the filament designs, the atoms immediately cool and severe interferences may be encountered. In the Massmann furnace, the atoms form as soon as the temperature of the tube wall reaches the **atomization temperature**. The gas within the tube will be somewhat cooler. L'vov has suggested a simple device that gives many of the advantages of isothermal atomization while essentially retaining the simplicity of the Massmann design. The sample is pipetted on to a small graphite **platform** only loosely connected to the tube walls. This platform is then heated, partly by radiative and convective means, and atomization occurs only when the **surrounding gas** is relatively hot. This device is sometimes referred to as the **L'vov platform** and miniature graphite plates can be purchased for this purpose. Alternatively, an old tube may be broken to provide fractions of the tube about one quarter of the circumference wide and 5 mm long. Using such platforms, some **interference effects** can be noticeably reduced.

ELECTROTHERMAL ATOMIZATION

Q. Why is rapid heating of the furnace to be preferred?

Q. Why is a feed-back device using a temperature sensor a better way of ensuring reproducible operation than is a simple applied power control?

Q. Why are contemporary furnace tubes manufactured so as to be shorter and smaller than their predecessors?

Q. To what are the advantages of so-called 'isothermal atomization' attributed?

11.5 ATOMIZATION MECHANISMS

An investigation of both thermodynamic and kinetic considerations is necessary in the understanding of atomization in graphite atomizers.

11.5.1 Thermodynamic considerations

Several possible reactions may be involved.

(i) Conversion of **metal salts to the oxide**. When heated strongly after deposition from aqueous solution, nitrates, sulphates and some chlorides are usually converted to the oxide.

(ii) **Evaporation of the metal oxide or metal halide** prior to atomization. Most **metal halides** are **volatile** and evaporate before atomization. Some metal oxides have **measurable vapour pressures** at the temperatures at which atomization is first observed to occur (the so-called **'appearance temperature'**). Using the simple **gas law**, 1 ng of a metal oxide (molecular weight = 100), completely vaporized into a volume of 100 mm^3 at 1000 K, would exert a vapour pressure of 6×10^{-3} mm Hg. Thus, unless this exceeds the saturated vapour pressure at 1000 K, complete evaporation could be expected.

(iii) **Thermal dissociation of the salt** or oxide. This will be related to the **temperature, pressure** and **other species** present. For dissociation of an oxide, the relevant equations are

$$-\Delta G = RT \ln K_p$$

and

$$\alpha = \frac{K_p}{K_p + \sqrt{P_{O_2}}}$$

where ΔG is the free energy of the reaction, α is the degree of dissociation of the oxide, K_p is the equilibrium constant for the dissociation and P_{O_2} is the partial pressure of oxygen. The partial pressure of oxygen is effectively controlled by the equilibria:

$$2C + O_2 \leftrightarrows 2CO$$
$$2CO + O_2 \leftrightarrows 2CO_2$$

(iv) **Reduction of the metal oxide**. Data concerning **carbon reduction** of metal oxides are readily available in forms such as the **Ellingham diagrams** used by chemists and metallurgists. The essential reactions to consider are

$$MO(s/l) + C(s) \leftrightarrows M(g) + CO(g)$$
$$C(s) + O_2(g) \leftrightarrows CO_2(g)$$
$$2C(s) + O_2(g) \leftrightarrows 2CO(g)$$

Such a thermodynamic approach can be extended to consider problems such as **carbide formation**:

$$MO + 2C \leftrightarrows MC + CO$$

The weakness of this approach is that it deals with **equilibrium criteria**, whereas the situation in a furnace and certainly on a filament is highly dynamic. It must also assume some dissociation at all temperatures, and thus the appearance temperature becomes that at which the free metal is first detectable: thus the parameter should be dependent upon the detection limit and concentration. Useful insights have been afforded by the application of thermodynamics, but clearly kinetic factors must also play a role.

11.5.2 Kinetic considerations

As L'vov first pointed out, to achieve analytically useful sensitivity, the **rate of formation** of the free atoms must be equal to or greater than their rate of **removal** from the atom cell.

$$\frac{dN(t)}{dt} = \left(\frac{dN}{dt}\right)_{formation} - \left(\frac{dN}{dt}\right)_{removal}$$

If N is the number of atoms at time t, dN/dt is the rate of change of the number of atoms.

In a graphite atomizer, the atoms will appear according to a kinetic **rate equation** which will probably contain an exponential function. As the number of atoms in the atom cell increases, so does the rate of removal, until, at the **absorption maximum** (peak height measurement), the rate of formation equals the rate of removal. Thereafter, removal **dominates**.

The response function, which is normally a **peak** and may be distorted to some extent by the electronics, clearly is the **difference** between the forma-

ELECTROTHERMAL ATOMIZATION

tion and removal functions at that time. Atoms leave the atom cell partly by **diffusion** and according to the velocity of the **purge gas**. The rate of formation of atoms is more difficult to identify.

L'vov first developed a kinetic model for atomization, based on **increasing temperature**, such as found in a rod-type system. Fuller developed a model for atomization under **isothermal conditions**, applicable to less-volatile elements in tube-type systems. Fuller's model assumes **first-order kinetics** and involves a number of other assumptions, but its usefulness has been demonstrated. For example, it confirms the usefulness of **integration** when atomization is slow (at relatively low temperatures or when investigating involatile elements), the enhancement of sensitivity available from **stopping the flow** of purge gas during the atomization cycle, and indicates methods for the control of interferences. More details of this model are given by Fuller (see Appendix C).

It is clear from experimental data that the rate of removal of the analyte can **exceed** the rate of supply. Thus, there is an advantage to be obtained in rapid heating (e.g. 1000 K s^{-1}) and stopping the purge-gas flow during atomization.

While a full treatment of kinetic theories is beyond the scope of this book, it is clear that they have added much to our understanding of observed events in tube furnaces. It can be expected that more sophisticated models will offer more comprehensive explanations of observed behaviour.

Q. Would you expect the electrothermal atomization mechanism for zinc to differ depending on whether the sample was dissolved in nitric or hydrochloric acid?

Q. How can we explain the observed shapes of peaks obtained using furnace atomizers?

11.6 INTERFERENCES

Electrothermal atomizers are usually regarded as **more prone** to interference than flames, although it has now been clearly demonstrated that the nature and extent of interferences may **vary** in different kinds of atomizers. We will now consider the problems encountered in the furnace type atomizers used in AAS.

11.6.1 Physical interferences

The steep thermal gradient means that any **variation in the sample position** (e.g. because of pipetting, surface tension or viscosity) will alter the atomization peak shape. **Peak area** integration will help to minimize this problem, as will a rapid-heating ramp and isothermal operation.

11.6.2 Background absorption

As already noted, the effects are usually severe. Large amounts of matrix are volatilized in a confined space. **Molecules** may exhibit **absorption spectra** in the region of interest; this is especially true of **alkali-metal halides**. Particulate **smoke** also contributes to this problem. **Light emission** from the incandescent walls may further distort the baseline. Every effort should be made to reduce background absorption effects in method development, e.g. by attempting to reduce matrix during the ashing stage. **Background correction** should be routinely applied (see section 4.2.3), remembering that background absorption is often at the 90% level.

11.6.3 Memory effects

Incomplete atomization of involatile elements can sometimes cause a problem. Often this can be overcome by firing a so-called 'cleaning' cycle at maximum power between analytical cycles.

11.6.4 Chemical interferences

(i) Losses of analyte as a volatile salt

This is particularly likely to occur when halides are present, at the ashing or atomizing stage at temperatures too low to afford atomization, and can lead to losses of, for example, $CaCl_2$ or $PbCl_2$. The use of hydrochloric acid for sample dissolution should be avoided. If chloride is present, excess nitric acid can be added to the sample and hydrogen chloride **boiled off** during atomization. Preferably, 50% ammonium nitrate can be added (with care) to give the reaction

$$NaCl + NH_4NO_3 \rightarrow NaNO_3 + NH_4Cl$$

| boils at | decomposes | decomposes | sublimes |
| 1413 °C | at 210 °C | at 380 °C | at 335 °C |

The compounds of Group V elements are often volatile, and loss of, for example, **arsenic, selenium and tellurium** during ashing of the sample can be reduced by the addition of **nickel ions**, to form nickel arsenide and such compounds. Other reported stabilization procedures, often referred to as **matrix modification** include: the stabilization of **cadmium** (normally lost at 500 °C) by 2% ammonium **phosphate** up to 1000 °C; the stabilization of **mercury** by **sulphide**; the reported prevention of **phosphorus** loss by **lanth-**

anum. Such modifications not only reduce losses on ashing, but also permit the use of **higher ashing temperatures** and thus may allow background and other matrix effects to be minimized.

(ii) Anion and cation interferences

Many of these have been reported, e.g. 0.1% (v/v) mineral acids interfere with several elements. While some insights are being gained in terms of the theories discussed above, interferences are usually combatted by matching standards and samples or by the method of standard additions.

(iii) Carbide formation

The apparent slow atomization of some elements may be caused by carbide formation. Rapid heating and a reproducible surface (e.g. a pyrolytic surface) help reduce the problem, as does the coating of the tube, e.g. with lanthanum using lanthanum nitrate solution.

(iv) Condensation

Some interferences appear to be **vapour phase effects** and are presumably due to occlusion of analyte elements into particles of matrix. The use of **platforms** in furnaces (see above), the use of **reactive purge gases** (e.g. hydrogen) and **dispersion** of the matrix, e.g. by using an organic acid such as ascorbic acid, can in some cases reduce such interferences.

Q. How can physical interferences be minimized?

Q. Is background correction more essential with flame or with electrothermal atomizers?

Q. How can interferences from chloride ions be minimized?

Q. How can arsenic, cadmium and mercury losses during the ashing stage be minimized, yet temperatures high enough to reduce the matrix be used?

Q. Suggest ways in which problems from the formation of thermally stable compounds may be minimized.

11.7 APPLICATIONS

Electrothermal atomization is particularly useful when the amount of sample is very **small**, when very **low levels** of detection are required and when the matrix is **dilute** or **volatile**. These criteria often apply to **clinical samples** (a pin-prick sample of blood produces only 50–100 mm^3 of whole blood, but this is sufficient for analysis using an electrothermal atomizer, thus it is not essential for an intra-venous sample to be taken). For such samples, often sample pre-treatment is not required, and body fluids and biological tissues can be **ashed *in situ*** in the furnace. This also applies to some **foods**, although others may need some preliminary **wet ashing**.

Oils can be injected **directly** or in a **dilute** form, e.g. diluted with xylene. Organometallic standards are recommended.

Metallurgical samples are perhaps not as amenable to electrothermal atomization as some types of sample, but as high sensitivity is often only required for the most-volatile elements, we can obtain useful information. Problems may be encountered from chloride when using aqua regia to dissolve samples. An interesting growth area is the placing of weighed **solid samples** directly into the furnace for ultra-trace analysis of volatile elements.

Waters are the subject of a voluminous literature, and various methods have been proposed to overcome some of the interferences encountered, e.g. adding ascorbic acid or lanthanum to remove interferences when determining lead in hard water. **Saline waters** present particular problems (e.g. from background absorption), and a preliminary separation may be advisable.

Air particulates are usually dissolved before analysis, but again solid samples (e.g. on glass fibre filters) have been analysed directly in furnaces.

The **standard addition** method of **calibration** (see Chapter 7) is often used to combat the uncertainties of varying interference effects in electrothermal atomization. However, care should be taken with this approach, as errors from spurious **blanks** and **background** may go undetected.

The **literature** on applications of electrothermal atomizers is still growing very rapidly and, because of the details of furnace programmes used, is well worth consulting. The applications tables in the Annual Reports on Analytical Atomic Spectroscopy (see Appendix C) offer the best way of accessing this information.

Q. Why is electrothermal atomization widely used in clinical applications?

Q. What are the advantages and disadvantages of the standard addition method for the above applications?

11.8 THE RELATIVE MERITS OF ELECTROTHERMAL ATOMIZATION

It is probably best to regard electrothermal and flame atomization as complementary techniques. Some factors which govern the choice of techniques for a given application are discussed below.

11.8.1 Advantages of electrothermal atomization

(i) **Increased sensitivity**: the **theoretical improvement** obtainable in **electrothermal atomization** has been calculated by several workers, including the author. Such calculations are based on the **poor nebulization** efficiency associated with flames (~10%), the **rapid dilution in the flame** with the expansion of flame gases, and the short residence time. Improvements in detection limits of furnaces compared with flames range up to 4000-fold for zinc and are typically in the range 100–1000-fold.

(ii) **Decreased sample size**: the minimum requirement of a flame is 500 mm^3, except where pulse nebulization is used (see section 8.2). For electrothermal atomization, sample sizes of 1–100 mm^3, typically 10–15 mm^3, are used. This means dramatic **improvements in absolute sensitivities** are obtained and measurements of picogram amounts of analyte are possible. Thus, electrothermal atomization offers analytical sensitivity comparable to that more often associated with neutron activation. Electrothermal atomization has particular advantages in situations where sample size is limited.

Table 11.1 shows typical concentrations and absolute limits of detection obtainable in electrothermal atomization, taken from a wide variety of the published literature.

(iii) *In situ* **sample treatment**: often tedious ashing procedures can be avoided by judicious choice of acids and ashing temperatures in the furnace.

(iv) Direct analysis of **solid samples**: solid samples can be placed directly in or on to electrothermal atomizers, often using purpose-built accessories.

(v) **Cheapness of operation**: there is a low consumption of argon, graphite tubes and electricity. Compare this with the consumption of gases by a flame.

(vi) **Safety** of operation: explosive gases and flames are avoided, less toxic fumes are produced, flame products are absent and smaller samples are used. Enclosed use means that radioactive samples can be handled.

(vii) Suitability for working in the **vacuum ultraviolet region of the spectrum**: argon does not absorb in the vacuum ultraviolet, whereas flame gases do.

Table 11.1 Selected absolute (pg) and relative (ng cm⁻³) detection limit data for various electrothermal atomizers used in AAS, AES and AFS

Element	Graphite furnace AAS pg	AAS ng cm⁻³ [a]	AES pg	AES ng cm⁻³ [b]	Graphite mini-furnace pg	AAS ng cm⁻³ [c]	Graphite rod AAS pg	AAS ng cm⁻³ [c]	AFS pg	AFS ng cm⁻³ [d]
Ag	0.1	0.001	700	14	0.2	0.04	20	4	1	1
Al	2	0.02	50	1	30	6.0				
As	6	0.06			100	20				
Au	10	0.1			10	2.0			4	4
Ba	50	0.5	30 000	600			20	4		
Be	0.1	0.001	200	4						
Bi	20	0.2	30 000	4.1	0.9	0.18			10	10
Ca	20	0.2			7	1.4				
Cd	0.1	0.001			0.3	0.06	0.08	0.016	0.15	0.15
Co	5	0.05	80 000	1600	0.1	0.02			20	20
Cr	10	0.1	500	10	6	1.2	8	1.6		
Cs			100	2.1	5	1.0				
Cu	2	0.02	900	18	20	4.0	5	1	1	1
Eu	40 000	400			7	1.4				
Fe	3	0.03	100	2.6	100	20	10	2	50	50
Ga	200	2	1000	23	3	0.6				
Hg	100	1	500	10	20	0.4				
K	1	0.01			100	20				
Li	5	0.05	0.08	0.0016	0.9	0.18	0.1	0.02	1	1
Mg	0.02	0.0002	4	0.07	5	1.0	2.5	0.5	5	5
Mn	0.2	0.002	60	1.1	0.06	0.012	20	4		
Mo	3	0.03	200	4.4	0.5	0.1				
Na	0.2	0.002			40	8.0				
Ni	10	0.1	0.1	0.0025	0.1	0.02	10	2	5	5
Pb	2	0.02	1000	23	10	2.0	4	0.8	10	10
Pd	20	0.2			5	1.0				
Pt	200	2			200	40				
Rb	10	0.1			200	40				
Sb	8	0.08			6	1.2				
Se	100	1	5	0.1	30	6.0			1000	1000
Si	50	0.5			100	20	100	20		
Sn	6	0.06			60	12				
Sr	5	0.05	60	1.2	5	1.0	75	15		
Ti	500	5								
Tl	10	0.1			3	0.6	50	10	50	50
V	100	1			100	20				
Zn [e]	0.05	0.0005			0.08	0.016	0.08	0.016	0.02	0.02

[a] 100 mm³ sample. [b] 50 mm³ sample. [c] 5 mm³ sample. [d] 1 mm³ sample. [e] Extrapolated limit in some instances.

(viii) **Unattended operation**: the use of an automatic sampler means that unattended, overnight operation is possible.

11.8.2 Disadvantages of electrothermal atomization

(i) **Time**: a typical programme cycle for electrothermal atomization may take 2 min, whereas a flame determination typically takes 15 s. In addition, the lack of a continuous reading makes setting-up more time consuming.
(ii) **Poor precision**: most imprecision in electrothermal atomization is associated with **manual pipetting**. Even when this possibility is removed by automatic sampling, it cannot be expected that discrete signals can offer precision similar to that of integrated continuous signals.
(iii) **Interferences**: current electrothermal atomizers still suffer from more interferences than the nitrous oxide–acetylene flame.
(iv) **Expense**: a good electrothermal atomizer with auto-sampler is an expensive accessory, costing perhaps half as much again as the original atomic absorption spectrometer (which probably will already be equipped for flame work).
(v) **Complicated programmes**: optimizing the conditions of electrothermal atomization is more complicated than in flame work.
(vi) **Small samples**: the small samples necessary in electrothermal atomization present problems in sample handling and with heterogeneity.

While it is to be expected that the effects of these disadvantages will continue to diminish as more becomes known about electrothermal atomization, currently it can be said that if there is sufficient sample for a flame analysis, and a flame cell offers sufficient sensitivity, it should be used. When flame sensitivity is insufficient, electrothermal atomization comes into its own, and is invaluable when either high sensitivity is required or only small amounts of sample are available.

Q. What are the particular advantages of electrothermal atomization?

Q. Would you use flame or electrothermal atomization for the following determinations: (i) zinc in a trade effluent at the 0.1 ppm level; (ii) cadmium in blood at the ppb level; (iii) lead in steel at (a) the 0.1% (w/w) level and (b) the 0.001% (w/w) level?

APPENDIX A

REVISION QUESTIONS

1 What is meant by the following terms:
   (a) ground state;
   (b) resonance line;
   (c) excited state;
   (d) absorption line half-width;
   (e) self-absorption;
   (f) Maxwell–Boltzmann distribution;
   (g) atomic fluorescence?

2 Describe the factors which cause broadening of spectral lines. In atomic absorption spectroscopy, why is it preferable for the source line-width to be narrower than the absorption profile?
How can this be achieved?
What are the differing requirements for resolution in monochromators for atomic emission spectroscopy and for atomic absorption spectroscopy?

3 Explain the operation of:
   (a) a photomultiplier tube;
   (b) a pneumatic nebulizer;
   (c) an electrothermal atomizer;
   (d) a hollow cathode lamp.

4 Describe the principles of:
   (i) atomic emission spectroscopy;
   (ii) atomic absorption spectroscopy.
Discuss the instrumental requirements for these two techniques.

5 (a) Discuss the importance of flame temperature in atomic emission spectroscopy and in atomic absorption spectroscopy.
   (b) Describe how instrumental requirements for obtaining optimum precision and accuracy may differ in flame atomic emission and in atomic absorption spectrometry.

6 With regard to both the theoretical principles and practical considerations of atomic absorption spectroscopy, discuss the design of **two** of the following:

APPENDIX A

   (a) nebulizer and burner systems;
   (b) light sources;
   (c) monochromators and detector systems.

7 Describe suitable instrumentation for sensitive analytical measurements in atomic absorption spectroscopy. Include a discussion of the ways in which the atomic population in the atom cell may be maximized and why the light source is always a line source.

8 In analytical flame spectroscopy, how are atomic populations usually formed from solutions? In your answer, include an outline of the conventional apparatus and basic processes involved, and explain how the atomic population may be maximized.
Discuss the essential features of suitable sources for:
   (a) atomic absorption spectrometry;
   (b) atomic fluorescence spectrometry.

9 Discuss why the following are often preferred for practical atomic absorption spectroscopy:
   (a) a modulated narrow line source;
   (b) a Czerny–Turner monochromator;
   (c) a small graphite tube atomizer;
   (d) microprocessor controlled curvature correction.

10 Discuss the reasons for the following.
   (a) Compared with flame atomizers, electrothermal atomizers generally result in enhanced sensitivity for atomic absorption spectroscopy.
   (b) Increasing the intensity of the source increases the sensitivity in atomic fluorescence spectroscopy, but has relatively little effect in atomic absorption spectroscopy.
   (c) Despite only a small fraction of atoms being thermally excited by the flame, flame emission spectroscopy is often as sensitive as atomic absorption spectroscopy.
   (d) Metals dissolved in organic solvents generally show enhanced sensitivity compared with those in aqueous solution.

11 Describe a typical electrothermal atomizer for atomic absorption spectrometry.
Critically compare graphite furnaces, air–acetylene flames, and nitrous oxide–acetylene flames as atom cells for atomic absorption spectrometry.

12 'Flames provide the most useful atom cells for atomic absorption spectrometry.'
Critically discuss this statement with particular reference to the analysis of 'real' samples.

13 Discuss the relative advantages and disadvantages of:

(a) atomic absorption and atomic emission;
(b) flames and electrothermal atomizers;
(c) hydride generation.

14 Compare the relative advantages and disadvantages of using atomic emission spectrometry and atomic absorption spectrometry for trace metal analysis.
What are the major sources of error in these two techniques and how may such errors be minimized?

15 'There have been persistent claims that, of atomic absorption, atomic emission and atomic fluoresence spectrometry, one is superior to the other two as an analytical technique. There have been similar claims that one of the techniques provides better sensitivity or selectivity. These claims have no sound theoretical or experimental basis.' Discuss these statements critically.

16 Discuss in detail the origins and effects of interferences in flame atomic emission and flame atomic absorption spectrometry, and describe how they may be minimized or eliminated in practice. Explain why some of these interferences are common to both methods, whilst others are found only in atomic emission.
Illustrate your answer with suitable examples wherever possible.

17 Compare theories of atomization in atomic absorption spectrometry, considering both flame and heated graphite (electrothermal) atom cells. Discuss various kinds of interference problems encountered in these types of atomizer.

18 Discuss possible causes of:
(a) upward curvature of calibration graphs;
(b) downward curvature of calibration graphs,
in atomic absorption, emission and fluorescence.
Suggest ways in which correction can be made to eliminate possible errors from such curvature.

19 Discuss, with appropriate reference to basic theory, the reasons for **five** of the observations in analytical flame spectroscopy listed below.
(a) Sodium can be determined more sensitively by flame emission spectrometry than by flame atomic absorption spectrometry, whereas the reverse is true for zinc.
(b) In the determination of lead in an effluent containing large amounts of sodium salts, the apparent lead absorption signal at 217.0 nm was reduced when deuterium arc automatic background correction was used.
(c) The addition of lanthanum to a plant digest solution increased the absorption signal for calcium.

APPENDIX A

(d) Virtually no atomic absorption can be observed for zirconium in an air–acetylene flame, but good signals can be observed in the nitrous oxide–acetylene flame.

(e) Far fewer atomic spectral interferences are observed in atomic absorption spectroscopy than in atomic emission spectroscopy.

(f) The addition of potassium to a lithium solution will greatly increase both the lithium absorption and emission signals in an air–acetylene flame.

20 Outline methods for performing the following determinations (approximate levels of the analyte are given in parentheses):

(a) sodium in soil extracts (50 mg l^{-1});
(b) manganese in cast iron (0.1%);
(c) lead in petrol (300 mg l^{-1});
(d) arsenic in a trade effluent (0.1 mg l^{-1});
(e) magnesium in tap-water (20 mg l^{-1});
(f) cadmium in tap-water (1 ng l^{-1});
(g) lead in blood (30 µg l^{-1});
(h) bismuth in nickel alloys (1 µg g^{-1});
(i) iron in silicate rocks (0.1%);
(j) chromium in welding fumes (1 µg m^{-3});
(k) mercury in seaweed (50 ng g^{-1});
(l) copper in beer (1 mg l^{-1}).

APPENDIX B
PRACTICAL EXERCISES

CALCULATIONS

A variety of units are used in practical atomic absorption work. The instrumental readings should preferably be recorded in units of absorbance (see section 4.1), or a simple multiple of absorbance. Percentage transmission must be converted to the logarithmic absorbance scale before such readings can be used. Concentrations are normally expressed as a weight or mass per volume, e.g. $\mu g\ cm^{-3} = \mu g\ ml^{-1} = mg\ l^{-1} = mg\ dm^{-3}$. Very often 1 $\mu g\ cm^{-3}$ is referred to as 1 ppm and so on. The use of this term, parts per million, should be discouraged, as it lacks rigour and is ambiguous. For example, if the lead content of a 1% solution of steel in aqua regia is reported as 1 ppm, does this mean 1 $\mu g\ cm^{-3}$ in the solution (i.e. 100 $\mu g\ g^{-1}$ in the solid) or 1 $\mu g\ g^{-1}$ in the solid? The term ppb, parts per billion, can cause further confusion because different definitions of 'billion' are used. Generally, the American billion, 10^9, is intended, i.e. 1 ppb = 1 ng cm^{-3} = 1 $\mu g\ l^{-1}$.

While auto-calibration facilities are provided on many instruments, plots of absorbance (y-axis) against concentration (x-axis) are still useful diagnostic tools. The exercise below presents typical data for such a calibration curve.

1 The following blank-corrected readings were obtained for the determination of nickel in steel, using nickel standards dissolved in iron solution (10g l^{-1}). The determination was performed by atomic absorption spectroscopy using an air–acetylene flame and the 232 nm nickel line.

Nickel concentration (mg l^{-1})	1	2	4	6	8	10	12
Absorbance	0.06	0.11	0.22	0.34	0.44	0.55	0.60

Plot the calibration curve for these data and comment upon any observed deviations from linearity.

Find the characteristic concentration (i.e. the concentration corresponding to an absorbance of 0.0044) for nickel (in iron solution) using this instrumentation by (a) extrapolation of the graph; (b) calculation from the slope.

If a 1% solution of steel gave an absorbance of 0.36, what would be the concentration of nickel in this solution and hence in the steel sample, as a w/w percentage?

(*Answers:* 0.08 mg l^{-1}, 6.5 mg l^{-1}, 0.065%)

2 The procedure for the method of standard additions was described in section 7.1. The following data were obtained for the determination of zinc in a plating works effluent by atomic absorption using an air–acetylene flame and the 213.9 nm zinc line.

	Analysis solution		
Effluent (cm^3) +	Distilled water (cm^3) +	10 µg cm^{-3} Zn standard (cm^3)	Absorbance
5	5	0	0.044
5	4	1	0.154
5	3	2	0.264
5	1	4	0.480

Plot the absorbance (*y*-axis) against the added zinc content of the solution (in µg; *x*-axis), extrapolate the graph and, from the negative *x*-intercept, calculate the concentration of zinc in the original effluent sample.

(*Answer:* 0.8 µg cm^{-3})

3 The calculation of the standard deviation of a series of readings can be used both to give an estimate of precision and to calculate the limit of detection (see section 4.2.5).

The following data were obtained for lead at the 217 nm line when spraying a standard solution of 1 µg cm^{-3} lead, using an atomic absorption spectrometer with background correction and a 2 s integration period. The instrument was auto-zeroed on the blank between readings.

Sample no.	1	2	3	4	5	6	7	8	9	10	11
Absorbance	0.051	0.043	0.048	0.057	0.055	0.051	0.044	0.049	0.055	0.059	0.059

(a) Calculate the mean absorbance.
(b) Calculate the characteristic concentration (for 1% absorption; 0.0044 absorbance).
(c) Calculate the standard deviation for the series of readings by calculating the difference between each reading and the mean, squaring each difference, summing the squares, dividing by the number of readings

AN INTRODUCTION TO ATOMIC ABSORPTION SPECTROSCOPY

minus 1 (i.e. 10) and calculating the square root. The use of exponentials (i.e. expressing the first difference as 9×10^{-4} absorbance units) will be essential to obtain a precise result using a typical hand-held calculator.

(d) Calculate the detection limit as the concentration which would apparently give a reading equal to twice the standard deviation.

(*Answers:* 0.0519 absorbance units, 0.085 $\mu g\ cm^{-3}$; 0.0056 $\mu g\ cm^{-3}$; 0.22 $\mu g\ cm^{-3}$).

Comment on possible reasons for the very noisy signals obtained and suggest how a lower detection limit might be obtained.

Before attempting any of the instrumental procedures outlined below, attention is drawn to the following comments about safety.

APPENDIX B

SAFETY PROCEDURES

Certain safety procedures must be observed when operating atomic absorption spectrometers. Manufacturers are under an obligation to outline these in the handbooks supplied with their instruments. A useful list of rules has been published by the Scientific Apparatus Makers' Association of the USA, and these are summarized below with some additional comments (for fuller details, see *Int. Lab.* May/June, p. 63 (1974)).

1 The laboratory should be well ventilated and the instrument provided with an adequate exhaust system, having air-tight joints on the discharge side; a typical recommended extraction flow for a flame is $3\,m^3\,min^{-1}$, with a hood of $40\,cm^2$. This is particularly important when spraying organic solvents containing chlorine or toxic metals, and when using the nitrous oxide–acetylene flame, as toxic oxides of nitrogen are formed. The fan should be located as close to the outlet as possible and the blades made of a heat-resistant material.

2 Gas cylinders must be securely fastened well away from any sources of heat or ignition, and preferably outside the laboratory in a purpose-built cylinder store. Cylinders and piping must be clearly marked to enable immediate identification of the contents.

3 After extinguishing the flame, all cylinders should be turned off at the cylinder valve and the gas lines vented to the exhaust.

4 The piping from the cylinders must be securely fixed so that it is unlikely to suffer damage.

5 Periodic checks for leaks should be made, by applying soap solution to joints and seals.

6 Special precautions are required for acetylene and, in the UK, special regulations are issued by the Health and Safety Executive (the '21 Point Guide').
   (a) Cylinders must be placed in a secure, weatherproof, adequately ventilated and fire-resistant store.
   (b) Acetylene cylinders must be stored upright and securely fastened.
   (c) A flashback arrestor should be fitted close to the cylinder.
   (d) Only approved regulators, connectors and piping should be used and these must be checked periodically to ensure that they meet the standards specified.
   (e) To prevent the formation of copper acetylide, copper tubing or fittings must never be used. Brass fittings should contain less than 65% copper.
   (f) In the UK, the acetylene pressure is required to be kept below 9 psi, unless special permission is obtained. Generally, acetylene should

never be run at a pressure higher than 15 psi (103 kN m^{-2}) because spontaneous explosions may result.

(g) Avoid gaseous acetylene coming into contact with copper, mercury, silver or chlorine.

(h) Never run an acetylene cylinder after the cylinder pressure has dropped to below 50 psi because the solvent (often acetone) will bleed from the cylinder. (In any case, below 100 psi, some deterioration in analytical performance may be observed.)

7   A nitrous oxide cylinder should not be used at a cylinder pressure of below 100 psi (6860 kN m^{-2}).

8   Before lighting the flame, always ensure that the water-trap beneath the burner/nebulizer chamber is full.

9   Take particular care when using flammable organic solvents. Use only the minimum quantities required. Keep surplus solvent well away, preferably in a container with a loose cover having a small hole for the uptake capillary.

10   Never directly view the flame or an electrothermal atomizer during an atomization stage, or a light source. Always wear protective eye wear, e.g. tinted safety spectacles.

11   Never leave a flame unattended.

12   Be aware of the electrical hazards of the instrument, particularly if an electrothermal atomizer is water cooled.

13   Never directly light or extinguish a nitrous oxide–acetylene flame, but only via a very fuel-rich air–acetylene flame. If appreciable carbon build-up occurs along the burner slot, switch such a flame off, via air–acetylene. Such deposits should only be scraped off in the absence of a flame.

14   As a general rule, when lighting a flame, turn on the oxidant first; when extinguishing a flame, turn off the fuel first. If the flame appears to start to fail, turn up the gases (not down, as this could cause a flash-back).

EXPERIMENTS

Experiment A. Basic operation and optimization of an atomic absorption spectrometer

For a given spectrometer, it will be necessary to refer to the manufacturer's handbook and instructions for operation.

1 Switch on the spectrometer.

2 Check that the drain tube is full of water, and that an air–acetylene burner is fitted.

3 Insert a magnesium hollow cathode lamp.

4 Set the lamp current to the recommended value.

5 Set the monochromator slit width to that recommended for magnesium by the manufacturer.

6 Turn the monochromator wavelength setting to 285 nm and carefully tune into the line maximum at 285.2 nm, adjusting any gain controls as necessary.

7 Check that the light from the lamp is apparently passing over the slot of the burner, 5–10 mm above the top of the burner, by using a white card.

8 Switch on the air compressor, check that the pressure is correctly set and the flow rate is set to that recommended by the manufacturer.

9 Switch on the acetylene cylinder, again checking the pressure and flow rate.

10 Ignite the flame and check that it is blue with a very faint tinge of yellow above the primary cone. If it appears lean or too fuel-rich, switch off and check the flow rates, adjusting as necessary.

11 After ignition, aspirate water continuously.

12 Using a 10 cm^3 measuring cylinder, check that the uptake rate of the nebulizer is within the range quoted by the manufacturer. Adjust if necessary.

13 Using pipettes and graduated flasks, prepare a 0.2 µg cm^{-3} standard solution from a 1 mg cm^{-3} stock magnesium solution.

14 Aspirate this solution. It should give an absorbance of about 0.2.

15 *Effect of hollow cathode lamp current.* Vary the hollow cathode lamp current, in steps of 1 mA, to a maximum of 5 mA either side of the recommended value. Measure the absorbance at each setting in an ascending and a descending mode. Remember to zero the instrument at each setting and if necessary adjust the gain before each reading. Plot a graph of absorbance against lamp current. Select the lamp current which gives the greatest ab-

sorbance, ensuring that the noise at that current is acceptable (you are really interested in the best signal-to-noise ratio).

16 *Effect of burner height adjustment.* Raise the burner until it just begins to obscure part of the light beam. Rezero the instrument and measure the absorbance of the magnesium solution. Lower the burner by 2 mm and repeat. Plot a graph of absorbance against burner height. Select the burner height setting which gives the greatest absorbance. Estimate where the light is now passing through the flame in relation to the primary reaction zone.

17 *Effect of varying the fuel–air ratio.* Vary the fuel–air ratio by increasing the fuel flow in steps until a strongly yellow flame (fuel-rich) is obtained. Again, adjust the zero at each setting and plot the absorbance against fuel flow. Identify the optimum fuel–air ratio.

18 *Adjusting the nebulizer.* If the instrument has an adjustable nebulizer or impact bead, this may be varied in a similar manner to identify the setting for the greatest absorbance signal.

19 *Effect of monochromator slit width.* For a range of slit width settings around the recommended value, determine the mean absorbance and standard deviation of a series of five readings of the 0.2 $\mu g\ cm^{-3}$ solution of magnesium. Calculate the characteristic concentration at each slit width and the limit of detection. Select the optimum slit width.

20 Prepare a series of magnesium standards of concentration 0.1, 0.2, 0.4, 0.6 and 0.8 $\mu g\ cm^{-3}$. Using the optimal settings (i) construct a calibration curve; (ii) calculate the characteristic concentration; (iii) identify the linear working range. Using the 0.1 $\mu g\ cm^{-3}$ solution, estimate the limit of detection of the method. Compare your results with the manufacturer's claims. If your instrument has a curve correction capability, use this to straighten the upper end of the calibration curve.

21 Mean levels of magnesium in adult urine are usually in the range 60–200 $\mu g\ cm^{-3}$. Thus, a two-hundred-fold dilution is necessary before aspiration. If either you or your boss can supply a sample, it might make an interesting conclusion to this experiment to determine magnesium in urine. Remember to sterilize all glassware afterwards if a real sample is used.

Experiment B. Optimization of graphite furnace operating conditions

1 Carefully read the manufacturer's instructions for the operation of the graphite furnace and the atomic absorption spectrometer.

2 Familiarize yourself with the correct operation of a 20 mm^3 micropipette.

3 Prepare a 0.2 $ng\ cm^{-3}$ solution of manganese nitrate.

APPENDIX B

4 With the furnace in position, a manganese hollow cathode lamp and background correction on, switch on the spectrometer. Ensure that the graphite tube is properly fitted into the furnace and that it is serviceable.

5 Pipette 20 mm^3 of sample into the furnace and set the furnace dry control to 95 °C for 1 min.

6 Start the furnace programme and observe the behaviour of the drop during the drying cycle ONLY, using a dentist's mirror.

7 Repeat this procedure at drying temperatures up to about 115 °C and select the 'dry' setting which allows the drop to evaporate smoothly, without spitting or frothing, in about 30 s.

8 Set this 'dry' setting with a time of 40 s.

9 Set the dry cycle as indicated, the ash cycle at a very low setting, e.g. 250 °C for 10 s, and the atomization temperature at 800 °C for 5 s. Set the tube-clean cycle to near maximum power for 5 s.

10 Again pipette 20 mm^3 of sample into the furnace.

11 Carry out the complete furnace programme, noting any absorbance peak (or area) during the atomization stage.

12 Repeat this procedure, setting the atomization temperature successively 100 °C higher. Plot a graph of peak (or area) reading against atomization temperature, until a plateau is reached, i.e. no further increase in absorbance is observed with increasing temperature for three temperature increments.

13 Set the atomization temperature to about 100 °C higher than the temperature of the onset of the plateau and increase the time to about 10 s.

14 Observe the profile of the peak during atomization and estimate the time when the manganese has substantially atomized, but any spurious blank peak is still minimal. Set this as the atomization line.

15 Optimize the ashing temperature. Initially, use an ash setting of 150 °C for 30 s, and the rest of the programme as decided above. Measure the absorption during the atomization stage. Increase the ash setting by 50 °C increments until a decrease is observed in the above absorption signal. (The plot of absorbance against ash temperature, the ash plot, will appear like a mirror image of the atomize plot.) Set the ash temperature to 100 °C less than the temperature at which the absorption signal begins to decrease.

16 Using the optimum programme thus defined: (i) construct a calibration curve for manganese (taking triplicate readings) and determine the linear working range, the reagent blank and characteristic concentration; (ii) estimate the limit of detection; (iii) determine the manganese content of the laboratory tap water by (a) direct interpolation from the calibration curve, and (b) the method of standard additions.

Compare your results for (i) and (ii) with the manufacturer's claims.
Compare the results obtained in (iii(a)) with those obtained in (iii(b)).

AN INTRODUCTION TO ATOMIC ABSORPTION SPECTROSCOPY

Experiment C. Determination of sodium in soil extracts by flame atomic emission spectrometry

Aim

To introduce the technique of atomic emission spectrometry and the effect of ionization suppression, and to illustrate a typical agricultural application.

Introduction

Sodium is an important element in soil analysis as it gives much information about the salinity and quality of a soil, and hence its agricultural utility. Available sodium is usually estimated by shaking a sample of soil with an ammonium chloride solution (1 mol dm^{-3} or 'normal'). This solution is used not for analytical utility, but in the hope that the fraction of the element extracted will have some agricultural significance. Therefore, the standards are also made up in ammonium chloride solution to correct for nebulization effects and possible blanks.

Sodium can be determined very sensitively by means of its intense atomic emission at 589 nm. Sodium is also readily ionized in the flame. This ionization can be suppressed by the addition of large concentrations of another easily ionized element (an ionization buffer). To overcome effects from differing amounts of such elements in different samples, an ionization buffer is included in all standards and samples.

The intensity of emission is exponentially related to flame temperature. It is therefore important that water is always sprayed into the flame to avoid excessive variations. AAS enjoys the advantage of being a ratio method and, as such, is easier to use. AES can be converted to a pseudo-ratio method by arranging for the strongest standard to read full-scale (or 90%) and periodically spraying this solution (solution 1), thus ensuring that flame conditions have not changed. The usual order is as follows:

deionized water; solution 1; water; solution 2; water; solution 3; water; solution 1; water; solution 4; water etc.

If significant variation is obtained whenever solution 1 is sprayed into the flame, conditions have altered. Corrections should be applied, either mathematically or with the gain control, to allow for these changes.

In sodium analysis, contamination is a severe problem. It is essential that all glass-ware is scrupulously washed free of detergents and that mouth or finger contact with solutions, via pipettes, beakers or the nebulizer capillary, is avoided. Test the extent of contamination problems by dipping your finger into a 100 cm^3 beaker containing 5 cm^3 deionized water. Report the sodium content of this contaminated water.

Provided the precautions outlined above are taken, AES offers a very simple and rapid method for soil analysis. Although the precision of AES is

APPENDIX B

probably not as high as that of AAS, in this application most of the errors are likely to be associated with sampling.

Apparatus

Atomic absorption spectrometer with flame emission facility
15 × 50 cm^3 standard flasks
100 cm^3 beaker
Pipettes and pipette fillers
Wash bottle
Graph paper

Reagents

Stock sodium solution (100 µg cm^{-3})
Stock potassium solution (10 000 µg cm^{-3})
Ammonium chloride solution (2 mol dm^{-3})
Soil extract

Procedure

Refer to the instrument instructions and familiarize yourself with the controls and operation of the instrument.

Before lighting the air–acetylene flame, ensure that you are familiar with the procedures for safely igniting and extinguishing the flame. Do not leave the flame unattended.

Use the manufacturer's recommended conditions for a 10 cm air–acetylene flame, a narrow slit width and 589 nm.

Ensure that the flame is blue, stoichiometric and without luminosity.

Spray water continually and allow a 10 min warm-up period. Follow the procedure for spraying solutions outlined above.

The effect of adding potassium to sodium solutions

Prepare, in 50 cm^3 standard flasks, seven solutions containing 10 µg cm^{-3} Na and, respectively, 1000, 500, 100, 50, 25, 10 and 0 µg cm^{-3} K. Prepare a blank solution containing only 1000 µg cm^{-3} K to check for sodium contamination. If necessary, correct for this.

Using the 10 µg cm^{-3} Na + 1000 µg cm^{-3} K solution, adjust the wavelength control to give maximum sodium emission. Adjust the gain controls and, if necessary, the slit width (see instruction manual) to give a reading of about 90 with this solution.

Spray the solutions as outlined above and record the intensity of emission in each case.

Plot the intensity of emission against the concentration of potassium.

AN INTRODUCTION TO ATOMIC ABSORPTION SPECTROSCOPY

The determination of sodium in a soil extract

Prepare, in 50 cm^3 standard flasks, a series of standards containing 1000 µg cm^{-3} K, 0.5 mol dm^{-3} NH$_4$Cl and, respectively, 0, 2, 5, 8, 10 µg cm^{-3} Na. Pipette 25 cm^3 of the unknown soil extract into each of two 50 cm^3 standard flasks, add 10 000 ppm K (5 cm^3) and make up to the mark.

Spray the standards and unknowns in order of their apparent concentrations using the procedure outlined. Note the emission readings.

Plot a calibration curve of emission intensity against sodium concentration.

Discussion

Discuss the shape of the graph showing the effect of adding potassium to sodium. What is the mechanism of this effect? Why is potassium added to all the solutions in the determination of sodium in the extract? Would you expect to see the same effects in atomic absorption and, if so, why?

What is the concentration of sodium in the soil extract? (Remember to correct for dilution.) What problems could arise if different batches of ammonium chloride were used in preparing the extracts and standards? What are the most probable sources of error in this experiment?

Does the flame AES method used appear to offer an acceptable level of precision and accuracy for the determination of sodium in soil extracts? What are the reasons for your conclusion?

Experiment D. A study of the interference effect of phosphate on calcium determinations by flame atomic emission spectroscopy

Introduction

In this experiment, the effect of phosphate on calcium during atomic emission is studied and the relative merits of three commonly proposed 'releasing agents' are investigated. At the end of the experiment, you will be asked to propose a method for the determination of calcium by atomic emission with minimum interference from phosphate.

It is suggested that the flame be lit and the instrument allowed to stabilize, spraying distilled water, for at least 10 min before use. This time could be usefully utilized reading the instrument manual. The usual order for spraying solutions into a flame is as follows:

distilled water; solution 1; water; solution 2; water; solution 3; water; solution 1; water; solution 4; water etc.

Solution 1 is sprayed into the flame periodically during a run to ensure that the flame conditions have not changed. If there is significant variation

APPENDIX B

in the readings obtained when solution 1 is sprayed into the flame, conditions have altered in the flame (for example, if the acetylene flow has changed). Corrections should be applied to every reading to allow for any changes in flame conditions. Solution 1 is usually the most concentrated solution or the one giving the largest reading.

Apparatus

Atomic absorption spectrometer with flame emission facility
250 cm^3 standard flask
20 × 50 cm^3 standard flasks
Pipettes and pipette fillers
Wash bottle
Graph paper

Reagents

Stock calcium solution (200 µg cm^{-3})
Stock phosphate solution (500 µg cm^{-3})
Stock EDTA solution (100 000 µg cm^{-3})
Stock lanthanum solution (100 000 µg cm^{-3})
Stock strontium solution (100 000 µg cm^{-3})

Procedure

The effect of adding phosphate to calcium solutions

Prepare a solution containing 100 µg cm^{-3} phosphate (250 cm^3) by appropriate dilution of the stock phosphate solution. Pipette 5 cm^3 stock calcium solution into each of a series of 50 cm^3 graduated flasks. From a burette, add the 100 µg cm^{-3} phosphate solution to each flask in increments so as to give solutions which, on dilution with distilled water, contain 0, 4, 8, 12, 16, 20, 30, 40 and 50 µg cm^{-3} phosphate. Each solution, when made up to the mark, will also contain 20 µg cm^{-3} calcium.

Scan the calcium emission spectrum from 420 to 430 nm. The monochromator should now be set at or near 422.7 nm (i.e. at the point of maximum intensity on the spectrum). In turn, spray the solutions into the flame, recording the intensity of the emission on the recorder in each case. Spray distilled water between each solution and allow time for the pen to return to the background level before the next solution is sprayed. Plot the intensity of calcium emission (deflection in divisions) against concentration of phosphate.

AN INTRODUCTION TO ATOMIC ABSORPTION SPECTROSCOPY

The effect of releasing agents

Before using the three releasing agents provided (EDTA, La and Sr), it is necessary to test that they are not contaminated by calcium. Thus, solutions containing 10 000 µg cm^{-3} releasing agent only should be sprayed first, in order to calculate any necessary corrections. If there is heavy contamination, that particular releasing agent need not be tested further.

After this test, take three series of 50 cm^3 graduated flasks and prepare solutions containing 20 µg cm^{-3} calcium and 10 000 µg cm^{-3} releasing agent (for the first series, use 5 cm^3 stock EDTA; for the second series, stock La (5 cm^3); and for the third series, stock Sr (5 cm^3)), and varying levels of phosphate. It is suggested that 0, 8, 16, 30 and 50 µg cm^{-3} phosphate be used first. Other levels of phosphate can be investigated later, if required.

Plot calcium emission against concentration of phosphate on the same graph, but label each plot clearly.

Discussion

In your discussion of this experiment, explain the shape of the graph showing interference, explain a probable mode of action of each releasing agent and suggest other parameters which might usefully have been investigated in this experiment had time allowed. Then answer the questions below.

1 Which releasing agent is to be preferred and why?

2 Outline a suitable method for the determination by atomic emission of calcium in phosphate-containing solutions.

3 Would you expect to see the same effects in atomic absorption and, if so, why?

4 What would you expect the effect of phosphate to be in calcium determinations using the hotter nitrous oxide–acetylene flame.

Experiment E. The determination of manganese in steel and cast iron by atomic absorption spectroscopy

A steel and cast-iron sample are required. Duplicate determinations of the manganese content should be made in each case.

Apparatus

Atomic absorption spectrometer
Manganese hollow cathode lamp
Pipettes and pipette fillers
Standard flasks
Wash bottle

APPENDIX B

Reagents

Stock manganese chloride solution (500 $\mu g\ cm^{-3}$)
Concentrated hydrochloric acid (Analar grade)
Concentrated nitric acid (Analar grade)

Instrument conditions

Use an air–acetylene flame and the manufacturer's recommended conditions for the manganese 279.5 nm line.

Procedure

Accurately weigh about 1.00 g of each of the samples and dissolve each in a mixture of concentrated hydrochloric acid (15 cm^3) and concentrated nitric acid (5 cm^3). This should be done in a fume cupboard, using gentle heat.

Filter each solution into a 100 cm^3 graduated flask and make up to the mark with distilled water. Pipette a 10 cm^3 aliquot of this solution into a 100 cm^3 graduated flask, and dilute to the mark with water. Spray this solution into the flame in the usual way and determine the absorbance.

Calculate the manganese content of the alloy by reading the manganese concentration in the solution from a calibration curve.

Preparation of calibration curve

Prepare 100 cm^3 of 50 $\mu g\ cm^{-3}$ manganese by suitable dilution of the stock manganese solution. From a burette, add 0, 2, 4, 6, 8 and 10 cm^3 of the 50 $\mu g\ cm^{-3}$ manganese solution to different 50 cm^3 graduated flasks, and dilute to the mark with water. In turn, spray each solution into the flame and construct a calibration curve of absorbance versus concentration of manganese.

It is suggested that the calibration solutions be prepared during the dissolution of the samples. All solutions can then be sprayed into the flame consecutively. As mentioned earlier, during the spraying of solutions into a flame over a period of time, the conditions in the flame might change. To check that conditions do in fact remain constant, it is suggested that the 10 $\mu g\ cm^{-3}$ manganese solution be sprayed regularly into the flame.

Experiment F. Background correction in atomic absorption spectroscopy

Aims

To investigate the interference of molecular absorption from sodium chloride on lead atomic absorption and the effect of background correction on this interference.

Introduction

It has become recognized that certain phenomena, such as scattering and molecular absorption, can cause non-specific or background absorption of light in the flame. Scattering and molecular absorption are wavelength dependent. Since they vary inversely with wavelength, the most acute problems are observed at low wavelengths. These effects are most severe in electrothermal atomizers, but are also observed in flames, e.g. when high concentrations of dissolved solids are being sprayed. Consequently, when elements with resonance lines of low wavelength are being determined in concentrated matrices, the possibility arises of spurious, elevated results.

This type of non-specific absorbance is the only significant form of spectral interference in atomic absorption and fortunately it can be greatly alleviated by 'background correction'. Using a typical monochromator, the amount of absorbance observed from an atomic line is negligible when using a continuum source, but molecular species give similar absorbances whether line or continuum sources are used. Thus, if light from a line source and from a continuum source are passed through the flame, it is not instrumentally difficult to arrange for the absorption using the continuum source to be subtracted from the absorption using the line source. A 'background corrector' is a system employing two such sources and automatically displaying the difference in the signals.

Apparatus

Atomic absorption spectrometer fitted with background correction facilities
$100\ cm^3$ standard flasks
Pipettes and pipette fillers
Wash bottle

Reagents

Stock sodium chloride ($50\ mg\ cm^{-3}$)
Stock lead nitrate solution ($100\ \mu g\ cm^{-3}$)

Instrument conditions

Use the manufacturer's recommended conditions for the determination of lead at 217.0 and 283.3 nm.

APPENDIX B

Procedure

Prepare a series of solutions (100 cm^3) containing 5 µg cm^{-3} Pb and, respectively, 0, 500, 1000, 5000 and 10 000 µg cm^{-3} of Na as NaCl.

Set up the instrument for lead, following the instrument instruction sheet or manual. Select the 283.3 nm Pb line, spray the 5 µg cm^{-3} lead solution and optimize conditions for lead absorbance.

Aspirate the samples prepared above and note the readings to two, preferably three, significant figures.

Change wavelength to 217.0 nm and repeat the readings taken above.

Switch on the background corrector.

Check that the energies of the two sources are approximately equal (this may involve reducing the Pb lamp current and auto-zeroing on water). Cease spraying water, and aspirate air. The background correction should compensate for the change in flame absorbance. Recommence spraying water.

Aspirate the samples again and note the 'corrected' readings.

Change the wavelength to 283.3 nm and repeat the exercise.

N.B. Are there any effects that can be expected from samples with high amounts of dissolved solids which tend to depress the observed signal?)

Return to the 217.0 nm wavelength, aspirate the 5 µg cm^{-3} Pb solution and take eleven readings. Aspirate the 5 µg cm^{-3} Pb + 10 000 µg cm^{-3} Na solution and take eleven readings. It is preferable to do this using a 2 s integrate mode.

Switch off the background correction mode and repeat these readings.

Calculate the standard deviation of the lead signals and lead with sodium chloride signals, with and without background correction.

Results

Report and discuss the significance of your results. Comment on the effectiveness of the background correction.

Include in your discussion answers to the following points.

(1) Is background correction equally necessary at both wavelengths? Why?
(2) At what level of lead would 10 000 µg cm^{-3} NaCl cause a 10% error (i) with, and (ii) without background correction?
(3) Can you explain any differences in the four standard deviations observed?
    Comment upon the implications of these obervations on the precision of likely analytical determinations.

APPENDIX C
REFERENCES AND BIBLIOGRAPHY

A list of useful texts and reference works is appended here. No attempt is made to give a selected bibliography of the many important original and review papers which have appeared. The advertising pages of *Analytical Chemistry* frequently contain readable introductory reviews of recent developments, and most of the journals below carry reviews from time to time.

JOURNALS

The following journals frequently publish original papers or reviews in this field.

Analyst
Analytica Chimica Acta
Analytical Chemistry (including the Annual Review issue)
Analytical Letters
Analytical Proceedings (of the Analytical Division of the Royal Society of Chemistry)
Applied Spectroscopy
Atomic Spectroscopy
Mikrochimica Acta
Spectrochimica Acta (Part B)
Talanta
Zeitschrift für Analytische Chemie

Additionally, there are many papers published in non-English language journals, application-oriented journals (e.g. *Clinical Chemistry*) and conference proceedings. *Analytical Abstracts* and *Chemical Abstracts* will cover many of these. There is also a new review journal, *Progress in Analytical Atomic Spectroscopy*.

Without doubt, the best tool for maintaining current awareness in this field (and that of electrical excitation) is *Annual Reports on Analytical Atomic Spectroscopy*, published by the Royal Society of Chemistry. Produced by a distinguished international editorial board, reports of all the above are critically assessed annually. Comprehensive **applications tables** enable practical workers to identify rapidly current methodology. *Annual Reports on Analytical Atomic Spectroscopy* is an invaluable tool for any analytical chemist involved in the practice of atomic spectroscopy.

APPENDIX C

BOOKS

These are listed alphabetically according to **author**, with a short comment where appropriate.

C. T. J. Alkemade and R. Herrmann, Fundamentals of Analytical Flame Spectroscopy, Adam Hilger, Bristol (1979). A useful account, with a sound theoretical basis.

M. D. Amos, P. A. Bennett, J. P. Matousek, et al., Basic Atomic Absorption Spectroscopy—a Modern Introduction, Varian (1975). A readable, light introduction for novices.

E. A. Angino and G. K. Billings, Atomic Absorption Spectroscopy in Geology, 2nd edn, Elsevier, New York (1972).

D. C. Burrell, Atomic Spectrometric Analysis of Heavy-Metal Pollutants in Water, Ann Arbor Science, Michigan (1974). A practical work-book.

G. D. Christian and F. J. Feldman, Atomic Absorption Spectroscopy: Applications in Agriculture, Biology and Medicine, Wiley–Interscience, New York (1970). Well-illustrated and readable, now in need of revision. Good balance of theory and application.

M. S. Cresser, Solvent Extraction in Flame Spectroscopic Analysis, Butterworth, London (1978). A useful compilation of methods.

L. De Galan, Analytical Spectrometry, Adam Hilger, London (1971). Necessarily a short, but readable introduction.

J. A. Dean and T. C. Rains, Flame Emission and Atomic Absorption Spectrometry, Vols. I–III, Marcel Dekker, New York (1969–1975). A comprehensive multi-author work, hence variable in quality, but excellent in places. Earlier volumes now in need of revision.

W. T. Elwell and J. A. F. Gidley, Atomic Absorption Spectrophotometry, 2nd edn, Pergamon Press, London (1966). A classic in its time, but obviously more up to date books are now available.

C. W. Fuller, Electrothermal Atomization for Atomic Absorption Spectrometry, Chemical Society, London (1977). The first text devoted to electrothermal atomization; the rapid growth in this area has tended to age this book rather rapidly.

A. G. Gaydon and H. G. Wolfard, Flames. Their Structure, Radiation and Temperature, 4th edn, Chapman & Hall, Andover (1979). See below.

A. G. Gaydon, The Spectroscopy of Flames, 2nd edn, Chapman & Hall, London (1974). The classic fundamental texts concerning the chemistry and spectroscopy of flames.

G. F. Kirkbright and M. Sargent, Atomic Absorption and Fluorescence Spectroscopy, Academic Press, London (1974). This tome is probably the most outstanding theoretical text of recent time.

T. Y. Komentani, Advances in Graphite Furnace Atomic Absorption Spectroscopy, Franklyn Institute Press, Philadelphia (1978).

B. V. L'vov, Atomic Absorption Spectrochemical Analysis, Adam Hilger, London (1970). A very useful text regarding theoretical aspects and the development of electrothermal methods.

R. Mavrodineanu and H. Boiteux, Flame Spectroscopy, Wiley, New York (1965). A useful source of information, particularly concerning basics and early developments.

R. J. W. McLaughlin, in Atomic Absorption Spectroscopy, Ed. J. Zussman, Academic Press, London (1977).

M. L. Parsons and P. M. McElfresh, Flame Spectroscopy: Atlas of Spectral Lines, Plenum Press, New York (1972). A tabulation of lines and bands of interest, useful to analytical flame spectroscopists.

AN INTRODUCTION TO ATOMIC ABSORPTION SPECTROSCOPY

M. L. Parsons, B. W. Smith and G. E. Bentley, *Handbook of Flame Spectroscopy*, Plenum Press, New York (1975). Contains much useful information on lines etc.

M. Pinta, *Atomic Absorption Spectrometry*, Adam Hilger, London (1974). A translation of a popular text written some years ago.

W. J. Price, *Analytical Atomic Absorption Spectrometry*, Heyden, London (1972). A very useful, practically oriented work, recently succeeded by the book below.

W. J. Price, *Spectrochemical Analysis by Atomic Absorption*, Heyden, London (1979). An excellent, practically oriented text with good accounts of instrumentation and application.

E. Pungor, *Flame Photometry Theory*, Van Nostrand (1967). Of historical interest.

J. Ramirez-Munoz, *Atomic Absorption Spectroscopy and Analysis by Atomic Absorption Flame Photometry*, Elsevier, Amsterdam (1968). An early text on atomic absorption.

J. W. Robinson, *Atomic Absorption Spectroscopy*, Dekker, New York (1966). One of the very earliest atomic absorption texts. Re-reading this now gives a good view of how the technique has developed since that time.

I. Rubeska and B. Moldan, *Atomic Absorption Spectrophotometry*, Iliffe, Prague (1969). A text which is no longer current.

B. L. Sharp, *Selected Annual Reviews of the Analytical Sciences, Vol. 4. High-Frequency Electrodeless Plasma Spectrometry*, Ed. L. S. Bark, Chemical Society, London (1976). As the title indicates, not entirely relevant to this book, but this review does include a very good account of atomic emission theory.

M. Slavin, *Atomic Absorption Spectroscopy*, 2nd edn, Wiley, New York (1978). A modern revision of a popular practical text.

V. Sychra, V. Svoboda and I. Rubeska, *Atomic Fluorescence Spectroscopy*, Van Nostrand Reinhold, London (1975). A comprehensive treatment of atomic fluorescence spectroscopy.

K. C. Thompson, *Atomic Absorption Spectroscopy, an essay review*, HMSO, London (1980). A practical introduction to the use of atomic absorption spectroscopy for the examination of waters and associated materials.

K. C. Thompson and R. J. Reynolds, *Atomic Absorption, Fluorescence and Flame Emission Spectroscopy*, 2nd edn, Charles Griffin, London (1978). A very practically oriented text, and a useful handbook for those involved in the laboratory.

J. C. Van Loon, *Analytical Atomic Absorption Spectroscopy: Selected Methods*, Academic Press, New York (1980). A useful laboratory handbook, especially for environmental work.

C. Veillon, *Handbook of Commercial Scientific Instruments, Vol. I. Atomic Absorption*, Marcel Dekker, New York (1972). A consumer guide to commercial equipment, now dated, and best updated using the instrument tables in *Annual Reports on Analytical Atomic Spectroscopy* (see above).

B. Welz, *Atomic Absorption Spectroscopy*, Verlag Chemie, Weinheim (1976). A popular German test translated into English.

J. D. Winefordner, *Spectrochemical Methods of Analysis, Vol. 9*, Wiley–Interscience, New York (1971). Includes a particularly good review of atomic fluorescence spectroscopic theory.

APPENDIX C

MANUFACTURERS' HANDBOOKS

Manufacturers of atomic absorption instrumentation are often useful sources of practical information. For example, the 'Perkin–Elmer Cookbook' is a compilation of analytical methods, the Instrumentation Laboratory 'flameless operations' handbook is very useful, Varian supply a series of booklets which are useful introductions to various areas of application and a series of booklets from Pye–Unicam introduces a wide range of theory and applications.

INDEX

A

AC spark, 35
Agricultural samples, 73
Aims, xiii
Air particulate analysis, 106
Alkemade, 1
Amalgamation, 86
Ammonium pyrrolidine
  dithiocarbamate, 76
Amplifiers, 10, 50, 55
Analytical utility, 77–79, 107–109
Anion interferences, 67–69, 80, 105,
  124–126
Annual Reports on Analytical Atomic
  Spectroscopy, 49, 73, 106, 130
Applications, 73–76, 106
Arcs, 34–35
Ashing, 64
Atomic absorption spectroscopy
  calibration curves, 55–56
  flames, 48–49
  instrumentation, 45–57
  line profiles, 24–26
  mercury, 87–88
  operation, 119–121
  origins and history, 1
  sources, 45–48
  theory, 42–45
Atomic emission spectroscopy
  calibration curves, 28, 29, 40
  instrumentation, 29–32
  line profiles, 24–26
  operation, 122–126
  origins and history, 1
  sources, 32–41
  theory, 27–29
Atomic fluorescence spectroscopy
  calibration curves, 58, 60
  flames, 61–62
  instrumentation, 59–62
  line profiles, 24–26
  mercury, 88–89
  origins and history, 1
  sources, 59–61
  theory, 58
Atomization mechanisms
  in flames, 16–20
  in electrothermal atomizers, 101–103
Automatic samplers, 83, 97, 109

B

Babington nebulizers, 40
Background correction, 32, 51–54, 67, 87,
  97, 104, 127–129
Barrier layer photocells, 30–31
Beenakker cavity, 38
Beer's law, 42
Biological samples, 73, 106
Blazed gratings, 5, 6, 7
Boling burner, 21
Boltzmann distribution law, 27–28
Books, 131–132
Branched capillaries, 81
Broadening, 24–26
Broida, 38
Buffers, ionization, 19, 67, 122–124
Bunsen, 1
Burners, 21–22, 23, 119

C

Calibration curves, 29, 40, 55, 58, 60,
  71–72, 79, 114–115
Carbide formation, 102, 105
Carbon filament atomizers, 94–95, 108
Carbon furnaces, 92–101
Carbon reduction, 102
Cation interferences, 67, 105
Characteristic concentration, 74–75, 91,
  108, 114–116

INDEX

Chemical interferences, 67–70, 104–105
Circular burners, 21
Clinical samples, 73, 106
Collisional broadening, 25
Compound formation interferences, 67–68, 80, 124–126
Concentric nebulizers, 16–18, 40
Condensation interferences, 105
Constant temperature furnaces, 99
Contamination, 63–64
Continuum sources, 43–45, 51–53, 59
Crops, 73
Cross-flow nebulizers, 40
Curve correction, 55, 71
Czerny–Turner, 6–7, 49

D

Dark current, 10
DC arc plasmas, 36–37
DC arcs, 34–35
Delves sampling cup, 82
Desolvation, 18
Detectability, see limits of detection
Detectors, 3, 8–10, 31
Deuterium arc, 51, 53
Diffraction gratings, 3, 5, 6, 7, 8, 49
Diffusion flames, 1, 62, 91, 99
Direct-line fluorescence, 58
Direct reading spectrometers, 35, 40, 76
Discrete sample nebulization, 81
Dissociation, 18–19
Doppler broadening, 24
Double beam instruments, 50–51
Drift, 65

E

Ebert, 6, 49
Echelle gratings, 6, 31, 37
Efficiency
  of atomization, 19, 80, 107
  of nebulization, 18, 80, 107
Effluents, 75, 115
Einstein transition probabilities, 27
Electrodeless discharge lamps, 47, 60–61
Electrolytic amalgamation for mercury, 86
Electrothermal atomizers, 91, 92–109, 120–121
Errors, 63–66
Evensen cavity, 38
Excitation in flames, 18–20

Excited state, 1
Expansion chambers, 17

F

Fertilizers, 73
Filament atomizers, 94–95, 96, 98
Filters, 4
Flame emission spectroscopy, see atomic emission spectroscopy
Flames
  advantages, 83–84
  air–acetylene, 14, 15, 22, 48, 62, 119–120
  air–hydrogen, 14, 15
  air–propane, 14, 15
  atomic absorption, 48
  atomic fluorescence, 62
  background, 22, 32, 83, 84
  burning velocity, 14
  diffusion, 1, 62, 90, 91, 99
  disadvantages, 84
  gas expansion, 80
  gas mixtures, 14–15
  laminar, 12
  limitations, 80–81, 84
  nitrous oxide–acetylene, 14, 15, 22, 31, 48
  photometry, 1, 29–31
  pre-mixed, 11
  propagation, 11
  radicals, 11, 15, 18
  separated, 22, 62
  spectra, 22
  stability, 11
  structure, 11–12
  temperature, 12–15, 19, 28–29, 65, 77, 79
  zones, 12
Food samples, 73, 106
Fraunhofer, 1
Furnace atomizers, 92–101
Furnace linings, 95

G

Gas chromatography, microwave plasma detector, 38
Geochemical analysis, 75–76
Glow discharge lamp, 35
Graphite filament atomizers, 94–95, 108
Graphite furnace atomizers, 92–101
Grimm discharge lamp, 35
Grotrian diagrams, 29, 30
Ground state, 2

135

INDEX

H

Hazards, 84
High solids burners, 21
Historical introduction, 1
Hollow cathode lamps, 35–36, 45–46, 60, 119–120
Holographic gratings, 5–6
Holtsmark broadening, 25
Huygens, 6
Hydride generation, 90–91
Hyperfine structure, 25

I

Impact bead, 17
Inductively coupled plasma, 38–40
Injection cup technique, 81
Instrumentation
   atomic absorption, 45–57
   atomic fluorescence, 59–62
   background correction, 51–54
   basic, 3–4
   detectors, 8–10
   electrothermal atomization, 92–101
   flame emission, 31
   hydride generation, 91
   mercury determination, 87–89
   optics, 4–8
   read-out, 3–4, 10
Integration, 55
Inter-conal zone, flames, 11, 12
Interference filters, 4, 31, 62
Interferences, 63–70, 90, 100, 103–105, 122–124
Ionization
   in flames, 19, 20, 67, 71, 80–81
   in hollow cathode lamps, 46
   interferences, 36–37, 67, 71, 122–124
Isothermal operation of electrothermal atomizers, 100, 103

J

Journals, 130

K

Kahn sampling boat, 82
Kinetic considerations of electrothermal atomization, 102–103
King furnace, 92
Kirchhoff, 1

L

Lambert's law, 42
Lasers, 60
Limit of detection, 56–57, 74–75, 77–78, 107, 108, 115–116
Line profiles, 24–26, 66
Line reversal method for temperature measurement, 12
Line sources, 43–48, 58, 59
Littrow, 5, 6, 49
Lock and key effect, 43
Lorentz broadening, 25
Lunegardh, 1
L'vov furnace, 92–94, 99
L'vov platform, 100

M

Massmann furnace, 93, 94
Matrix modification, 104–105
Maxwell–Boltzmann distribution, 13, 24, 27–28
Meker burner, 20–21
Memory effects, 104
Mercury
   absorption cells, 87–88
   determination, 86–89
   fluorescence cells, 88–89
Metals analysis, 76, 106, 114–115, 126–127
4-Methylpentan-2-one (MIBK), 76
Microprocessors, 55, 71
Microsampling cup, 81
Microwave plasmas, 37–38
Mineral samples, 75–76
Modulation, 10, 50, 62
Molecular absorption, 51, 66–67
Monochromators, 4–5, 40, 50, 62, 120
Multi-element lamps, 46

N

Natural broadening, 24
Nebulization
   pneumatic, 16–20, 39–40, 107
   pulse, 81
   ultrasonic, 18, 39–40
Newton, 1
Non-radiative transition, 2
Noise, 10, 47, 48, 53, 55, 62, 65, 77

INDEX

O

Occlusion interferences, 69–70
Oils, 73, 106
Operator errors, 64–65
Optics, 4–8
Organic samples, 73, 106
Organic solvents, 18, 73, 75, 106
Organic standards, 73
Oxide band emission, 27

P

Petrochemicals, 73, 106
Photomultiplier tubes, 3, 8–10
Planck's equation, 2
Plasma sources, 36–41
Platform atomization, 100
Pneumatic nebulization, 16–20, 39–40, 107
Polychromators, 32–33
Precision, 55–56, 78–79, 107, 115
Pre-heating zone, flames, 11, 12
Prior knowledge, xiii
Primary reaction zone, flames, 11, 12
Prisms, 5
Protective agents, 69, 124–126
Pulse nebulization, 81
Pyrolytic graphite, 92, 95

Q

Quenching, 61, 84

R

Random access memories, 56
Read only memories, 56
Read-out systems, 3, 10, 55–56
Reduction–aeration method for mercury, 86–87
Refractory compound interferences, 69
Refractor plate, 32
Releasing agents, 69, 124–126
Resolution, 31, 49, 66
Resonance broadening, 25
Resonance fluorescence, 58–59
Resonance lines, 24
Response curves, photomultipliers, 9

S

Safety, 11, 84, 107, 117–118
Sample pre-treatment, 63–64, 107
Sample size, 81–82, 107, 109
Sampling boat, 82
Saturation fluorescence, 61
Scattering, 58, 84, 104
Sea-water analysis, 106
Secondary reaction zone, flames, 11
Self-absorption, 25–26, 29
Self-reversal, 26
Sensitivity, 56–57, 74–75, 77–78, 107–109
Silicate analysis, 75–76
Simultaneous multi-element analysis, 32–34, 79
Slot burners, 21, 48
Sodium borohydride reduction, 87, 90
Soil samples, 73, 122–124
Solar-blind photomultiplier tubes, 10
Solid phase interferences, 67
Solid samples, 33–36, 81, 82, 97, 99, 106, 107
Solvent extraction, 76
Sparks, 35
Spectral interferences, 66–67, 77, 79
Sputtering, 35, 46
Standard additions, 71–72, 106, 124–126
Standard deviation, 57, 115–116
Standards, 64–65, 71–72
Stark broadening, 25
Study time-table, xiii
Suppressor, ionization, 19, 67, 122–124

T

Tantalum filament atomizer, 98
Temperature
 control, 99
 flames, 12–14
 feedback, 100
 measurement, 12–14, 100
 plasmas, 36, 38, 39
 programmes, 97, 109, 120–121
Thermodynamic considerations of electrothermal atomization, 101–102
Tin(II) chloride reduction, 86
Tube furnaces, 94–98
Tubes in flames, 81, 91
Two-line method for temperature measurement, 13

U

Ultrasonic nebulization, 18, 39–40, 80

INDEX

Units, 114
Uptake rate, 65

V

Vacuum ultraviolet region, 107
Venturi effect, 17
Visual display units, 56
Voight profile, 25

W

Walsh, 1
Waters, 75, 106
Wavelength dispersion, 3, 4–5
Wavelength modulation, 32, 34
Wavelength scanning, 32
West rod, 94–95
Winefordner, 2
Wollaston, 1
Woodriff furnace, 99
Working range, *see* calibration curves

X

Xenon arcs, 43, 59

Z

Zeeman
 broadening, 25
 effect, 52, 54